MATHEMATICAL MODELLING FOR INFORMATION TECHNOLOGY:
Telecommunication Transmission, Reception and Security

MATHEMATICS AND ITS APPLICATIONS
Series Editor: G. M. BELL, Professor of Mathematics,
King's College London (KQC), University of London

STATISTICS AND OPERATIONAL RESEARCH
Editor: B. W. CONOLLY, Professor of Operational Research,
Queen Mary College, University of London

Mathematics and its applications are now awe-inspiring in their scope, variety and depth. Not only is there rapid growth in pure mathematics and its applications to the traditional fields of the physical sciences, engineering and statistics, but new fields of application are emerging in biology, ecology and social organization. The user of mathematics must assimilate subtle new techniques and also learn to handle the great power of the computer efficiently and economically.

The need for clear, concise and authoritative texts is thus greater than ever and our series will endeavour to supply this need. It aims to be comprehensive and yet flexible. Works surveying recent research will introduce new areas and up-to-date mathematical methods. Undergraduate texts on established topics will stimulate student interest by including applications relevant at the present day. The series will also include selected volumes of lecture notes which will enable certain important topics to be presented earlier than would otherwise be possible.

In all these ways it is hoped to render a valuable service to those who learn, teach, develop and use mathematics.

Mathematics and its Applications
Series Editor: G. M. BELL, Professor of Mathematics, King's College London
(KQC), University of London

Series continued at back of book

MATHEMATICAL MODELLING FOR INFORMATION TECHNOLOGY:
Telecommunication Transmission, Reception and Security

Editors:

A. O. MOSCARDINI, M.Sc., Ph.D.
Department of Mathematics and Computer Studies

and

E. H. ROBSON, M.Sc., Ph.D.
Dean of Science
both of Sunderland Polytechnic

ELLIS HORWOOD LIMITED
Publishers · Chichester

Halsted Press: a division of
JOHN WILEY & SONS
New York · Chichester · Brisbane · Toronto

First published in 1988 by
ELLIS HORWOOD LIMITED
Market Cross House, Cooper Street,
Chichester, West Sussex, PO19 1EB, England
The publisher's colophon is reproduced from James Gillison's drawing of the ancient Market Cross, Chichester.

Distributors:

Australia and New Zealand:
JACARANDA WILEY LIMITED
GPO Box 859, Brisbane, Queensland 4001, Australia

Canada:
JOHN WILEY & SONS CANADA LIMITED
22 Worcester Road, Rexdale, Ontario, Canada

Europe and Africa:
JOHN WILEY & SONS LIMITED
Baffins Lane, Chichester, West Sussex, England

North and South America and the rest of the world:
Halsted Press: a division of
JOHN WILEY & SONS
605 Third Avenue, New York, NY 10158, USA

South-East Asia
JOHN WILEY & SONS (SEA) PTE LIMITED
37 Jalan Pemimpin # 05–04
Block B, Union Industrial Building, Singapore 2057

Indian Subcontinent
WILEY EASTERN LIMITED
4835/24 Ansari Road
Daryaganj, New Delhi 110002, India

© 1988 A.O. Moscardini and E.H. Robson/Ellis Horwood Limited

British Library Cataloguing in Publication Data
Mathematical modelling for information technology:
Telecommunication transmission, reception and security. —
(Ellis Horwood series in mathematics and its applications. Statistics and operational research)
1. Telecommunication systems — Mathematical models
I. Moscardini, A.O.   II. Robson, E.H.
621.38   TK5102.5

Library of Congress CIP data available

ISBN 0–7458–0381–4 (Ellis Horwood Limited)
ISBN 0–470–21024–9 (Halsted Press)

Phototypeset in Times by Ellis Horwood Limited
Printed in Great Britain by Unwin Bros., Woking

Table of contents

12   Modelling workloads for local area network performance analysis
A. S. Wight

13   Network and coding courses at undergraduate and diploma level
M. G. Everett and F. M. Tyler

14   Reliability and security issues in distributed systems
J. E. Dobson

15   UK research on implications of information and communication technologies
W. H. Melody

Preface

The Chapters of this volume were originally papers presented orally at the tenth POLYMODEL conference entitled 'Modelling for the Communication Industries' held at Sunderland Polytechnic in May 1987. The conferences are organised annually by the North East of England Polytechnic's Mathematical Modelling and Computer Simulation Group — POLYMODEL. The Group is a non-profit-making organisation based on the mathematics departments of the three polytechnics in the region, and has membership drawn from those educational institutions and from regional industry. Its objective is to promote research and collaboration in mathematics and computer-based modelling. Clearly, since the increasing sophistication of communications technology is so dependent on mathematical and simulation methods and techniques, the conference was both appropriate to those seeking new areas in which to apply their mathematical skills and timely for those endeavouring to contribute to the technological developments in this field. This view was endorsed by Dixon (of the ESPRIT Directorate) who, in his opening address, indicated that the main theme of the conference would be included in the new EEC initiative, RACE, which aims to promote European research into communications technology; also by Melody (see Chapter 15) who outlined a programme, sponsored by the ESRC, to promote research in the United Kingdom into the economic and social impact of these developments. In addition, Thompson (Chapter 10) indicated a strong need for forums, such as that provided by POLYMODEL, where modelling ideas and methodologies can be discussed and related information disseminated. Thus this collection of papers should be of interest to a wide readership of mathematicians, computer scientists and engineers involved in the theories and practices relating to communication technology.

The volume is divided into four parts, each of which begins with a short introduction. The first five chapters make up the part on Signal Processing. The second part comprises four chapters in the relatively new area of Communication Security and Part 3 contains five chapters on Networks. The final part on the Social

Impact of Communication contains three chapters which reflects some social science views which are complementary to the (essentially natural science) ideas and theories propounded in the earlier parts.

The editors would like to take this opportunity to thank all of the participants of POLYMODEL 10 for making the conference so successful and particularly those who presented the papers. The editors are especially indebted to Peter Dunne, Barry Lewis, Ken Lever, Terry Wilkinson and Emrys Hughes who contributed in different ways to the publication of this volume. They are also grateful for the administrative support given by Juliet Hudson.

Sunderland Polytechnic   A. O. Moscardini
July 1987   E. H. Robson

Part 1:

Signal processing

Amongst engineering disciplines, Signal Processing may give the impression of having a mathematical basis wider and deeper than most. But this is not so: the vast majority of models in signal processing are either time-domain or frequency-domain in one form or another, and the subject consequently exhibits a certain degree of theoretical coherence. Despite an apparent diversity of subject matter, this coherence is manifested in the first group of five chapters, in an interesting way that could well carry an important message for those responsible for the mathematical training of young people intending to make their careers in the Communication Industries.

The first chapter in this part, presented by Mason, reviews compression algorithms for reducing the transmitted data rate of video images, using adaptive one- and two-dimensional time-domain coding techniques and their frequency-domain counterparts.

Ingleby's chapter addresses the signal recognition problem from the unusual viewpoint of a mathematical taxonomist, invoking mathematical models such as lattice representations of signals, resulting in noncommutative algebras requiring the von Neumann interpretation of probabilistic inference.

The third chapter, by Godwin et al., concentrates on feature modelling and the design of simplified signal and spectral processing algorithms for speech identification.

Spectral processing makes another appearance in the chapter by O'Reilly and Monteiro on error control coding for optical communication links. Conventional Fourier-based techniques, such as those discussed by Mason, employ the real or complex number field as the underlying mathematical model. In this application, however, a Galois field is used instead, permitting the error–correction decoding to be viewed as a form of spectral estimation.

The final chapter in this part, by Lever et al., outlines another use of lattices to represent signal processing architectures for adaptive equalisation of digital signal distorted by transmission over a fading radio channel. Once more a noncommutative algebra plays an important role in a design based on the Z-transform discrete-system counterpart of the Fourier model, and the emergence of associated Cyclic Group and

Galois field models underlines the importance of algebraic structure in designing fast algorithms by means of the 'divide-and-conquer' approach beloved by computer scientists.

In summary, a distinct harmony is detected in the ubiquitous use of signal representatives based on the structure of abstract algebra. This is not surprising; in this age of digital communication using discrete systems, it is surely time to acknowledge in mathematics education at all levels that discrete mathematics is the predominantly natural model.

1

Algorithmic techniques for image and video coding

J. R. M. Mason
GEC Research Ltd, Hirst Research Centre, East Lane, Wembley, Middlesex, HA9 7PP, United Kingdom

1.1 INTRODUCTION

With the increasing demand for large amounts of information to be transmitted as soon as possible and continuing developments in semiconductor technology, combining a steadily rising speed of processing with a steadily deceasing cost of components, interest in low bit-rate picture coding is currently running higher than ever before. There are several applications which have spurred on research and development work on coding, some of which are currently playing an important role in the business world and promise to expand that role into the home in the medium to long term. They include such topics as picture facsimile, picture archiving, remote sensing, surveillance systems and video-conferencing.

This chapter is a review of some of the techniques that have been proposed in order to produce compressed data describing a moving video sequence.

One question we should perhaps ask ourselves is 'Why do we need data compression of image sequences?'. With the advent of hardware enabling more sophisticated, and thus better, coding methods comes the approach of fibre optic communications and optical disks for transmission of data of the order of hundreds of megabits per second and storage of several gigabytes per disk. For most applications, transmission of images at hundreds of megabits/second is quite adequate, for at these rates, one is transmitting data at a rate almost equivalent to the rate that the best video cameras can capture. However, this is a fine argument provided that there are enough fibre optic links between all those people who wish to communicate. Clearly this is never likely to be the case, since international communication is a mode of contact which is currently an important business application and is likely to become more so in future years. Thus, satellite communications must be used. For a satellite to be used for communications, it must remain in geostationary orbit. There is a limit to the number of satellites which may do this, and each satellite's communications channel is band-limited. Therefore one is charged for the size of channel one requires and, at least currently, this is very expensive. In addition there are various bit-rates which are acceptable for different applications of video coding, e.g. 64–384 kbits for videophone, 384k–2 Mbits/s for video-conferencing, 4–32 Mbits/s for digital television and 32–140 Mbits/s for High Definition Television (HDTV). When current-plans to have an Integrated Broadband Communications network (IBCN) are implemented it will be of interest to be able to multiplex two or more of these types of application at a given bit-rate to supply the home with the information it desires.

To put some of these applications into perspective, a standard PAL television signal sampled according to CCIR recommendations consists of 576 lines, 720 pixels per line, 25 frames per second and requires about 120 Mbit/s in all. In order to send these data down a digital telephone line (which operates at 64 kbits/s) we therefore need the equivalent of about 2000 lines.

The viewphone is an idea that has been around for a while, although it has yet to be perfected. The ultimate aim is to produce a 64 kbits/s (56 kbit/s in the USA) video coder (codec) which will allow telephone calls to be supplemented with a moving visual image of the caller. There are currently some systems on sale in the USA. Although these are to be complimented on their early entry into the market, they are rather lacking in performance. The reduction of a factor of 2000 required to achieve this end at first sight appears daunting, but one should bear in mind that television is a very high quality signal and can be subjected to reduced resolution for this particular application, whilst remaining of acceptable quality (compare high fidelity sound systems with the telephone). It is expected that, if this product can be made at a low price, it will eventually find its way into the domestic market.

Video-conferencing is a current application in video coding and has the benefit of a European (2 M/bits/s) and a United States (1.5 Mbits/s) standard, allowing equipment made by different manufacturers to be connected as easily as equipment made by the same manufacturer. It is likely that there will be a new world standard for this at 384 kbits/s soon, which will have the effect of reducing operating costs substantially. The main market for this application is in the business world, where the need for many expensive (both in travel and in travel time) meetings of business colleagues based in different parts of the world can be removed by video-conferencing. Such codecs usually have a high resolution graphics mode and sometimes an encryption option for maintaining confidentiality for board-room-level meetings.

Electronic News Gathering is an application requiring video coding, for if a remote event can be captured with a TV camera, it should usually be sent as quickly as possible to the headquarters, either for subsequent home broadcasting, or perhaps for military or political purposes. This usually means satellite transmission which, as discussed above, is costly and thus the picture should be coded. In some instances of news gathering, the ability to transmit news 'as it happens' hugely outweighs any small losses in quality due to data reduction.

Entertainment TV — both the broadcast signal at the current resolution and the still-to-be-decided resolution of High Definition Television — will eventually be digitally transmitted, the reasons for this being greater flexibility for manipulating the image, better response to poor reception and compatibility with other data in IBC networks. Television companies have been using digital television for a few years now in order to get some special effects which cannot be easily achieved using analogue systems, although of course the signal must be translated into an analogue form for current transmission systems.

The basis of all coding techniques for image compression is the same. In order to appreciate the logic behind it one must realise the difference between the terms 'information' and 'data'. Information is a concept that has spawned a whole field of technical work, called Information Theory (started by Claude Shannon in 1949 [1]). The term can be defined as a quantity, which, if changed, renders a new meaning to the quantity or the object of which it forms part. The word 'data' has an intuitive meaning to most people, at its simplest being something that can be represented as a sequence of 1s or 0s. Thus, in image coding, an example of a piece of data that is not 'information' would be, say, the least significant bit of a pixel in an image: if we changed it there would be no perceived difference to the viewer.

We call those bits of data which are not information the redundant bits. It is the job of the codec to remove as much redundancy as possible, whilst transmitting faithfully the information content. Of course it is not possible to remove all the redundancy whilst leaving all the information, since the definition above means that we would require a model of human perception in order to function automatically. As a clue to the sort of data-rate for image compression that is theoretically possible, we can point to the belief that the human brain processes visual information at around 50 bits/s.

It is interesting to note that the aims of video compression are entirely contradictory to the philosophy of error robustness, in that compression requires that as much redundancy as possible be removed, whilst for a signal to be robust over a noisy channel, redundancy, is necessary. The requirements for encryption, however, are well satisfied by a well-compressed image sequence.

Before we attempt to describe compression algorithms, we should say something about the assessment of algorithms. This is a very difficult process. In order to get an objective measure of the performance of an algorithm, one must make measurements. However, all images are intended to be viewed by the human observer, and therefore must be processed by the human perceptual system, which is not sufficiently understood to be incoporated into a computer. Therefore it is not possible to make suitable objective measurements of an image. Having said this, many researchers use the 'mean-squared error' criterion for assessment, whilst acknowledging that it is far from perfect. For example, one can easily imagine a picture of a person who has had, say, an eye removed by the coder, thereby destroying a large amount of information. If the same picture is translated spatially by one pixel diagonally, the picture would still be visually perfect to the observer, but the sums of squares of the pixel-to-pixel differences would quite probably be higher in the latter than in the former. The only way to measure picture quality is subjectively.

1.2 ALGORITHMS FOR VIDEO CODING

There are several classifications of algorithms which have been proposed for video coding; we shall use two methods. The first, a generic taxonomy, will show mainly the way different algorithms may be applied to suit different applications, the second being a proper algorithmic classification.

There is a partitioning of the use of algorithms into three classes: Intraframe, Interframe, and Hybrid.

Intraframe refers to any method of coding which uses only the redundancy within single frames (i.e. spatial redundancy) to perform image compression. This method, although not using all redundancy available, and thus running generally at a higher bit-rate, has the advantage that it is able to code up sequences whose content has a higher degree of motion — either a single uniform motion as in a camera pan, or in multiple motion directions, but with a single global motion function (as in a zoom), or in multiple motion directions with no global motion function (a scene with many objects travelling at different velocities) — since by not using motion functions, it cannot be constrained by motion and thus allows any type of motion.

Interframe coding comprises, in contrast to intraframe coding, techniques which purely use temporal redundancy in sequences to perform data reduction, that is the coding is only based on changes to the image occurring from frame to frame. These techniques are particularly suited to those applications which will have a stationary camera, and usually movement occurring only in the foreground of the image rather than the background, thus ensuring that only a small proportion of the image moves from frame to frame. There are some algorithms, however, which have been adapted to work a largely uniform motion of the scene (pan, zoom, etc.).

Most codecs, however, rely on both spatial and temporal redundancy reduction, thus usually allowing greater compression; we call these methods 'hybrid techniques' (not to be confused with hybrid algorithms that we will discuss later).

We will now examine the algorithms in more detail. For the purposes of discussion, we shall discuss three algorithms, which can be classified as follows: spatial predictive, temporal predictive, and hybrid.

1.3   SPATIAL PREDICTIVE METHODS

The philosophy behind predictive methods is that when a prediction is made of the signal from previously transmitted portions of the signal, then provided the prediction is fairly sensible, the error in the prediction — which is the quantity that is coded — will have better statistics for coding purposes, for example the variance of the error signal will be smaller than the variance of the original signal.

In order to implement a predictive coder, one must also have a decoder within the coder. This is necessary in order to make sure that errors in quantisation are not propagated as coding continues. In other words, at any given time, the coder must know what data the decoder is about to process (apart from channel errors). Although this will mean extra circuitry at the coder, it has the advantage that the signal that will be received is readily available at the coder.

The typical algorithm using pedictive coding is DPCM, or Differential Pulse Code Modulation. In this situation, the picture value of picture element x_j is predicted from the values of pixels within a neighbourhood $N(x_j)$. So if $P(x_j) =$ picture value of pixel x_j, then

$$P(x_j) = \sum_{x_i \in N(x_j)} a_i \cdot P(x_i) + e(x_j)$$

where $e(x_j)$ is the error term which accepts that the picture is not totally predictable, and the a_i are coefficients for each of the spatially distinct neighbouring pixels. The values of the a_i can be determined in a training mode by taking a 'representative' image or sequence of images and by calculating the autocorrelation function, and choosing the a_i to give, by way of a least squares estimation, a minimal sum of errors. In a codec, the error signal only is sent, which, owing to its small variance, can be quantised and transmitted with fewer bits than was possible with the original signal for the same fidelity.

This technique is therefore modelling the signal with an all-pole filter, where the coefficients of the filter are determined *a priori*. This concept works well when the image is a fairly flat picture, with few edges. But at the edges, where the prediction error is large, the coder will suffer, for, if a variable length code is used, the code word will be of long length.

A further development of the DPCM technique is to make the error quantisation adaptive with the data currently being coded. This method is called Adaptive DPCM and (ADPCM) and can be achieved in one of two ways. Forward ADPCM entails isolating a block of pixels, and calculating for that block which of a set of error quantisers would be more efficient at coding the block, with respect to the mean-squared error. This would allow coarser quantisers in regions of high variance and finer ones in regions of low variance. Backwards ADPCM allows the quantiser to be changed on the basis of the received data. That is if, for example, the smallest step for the quantiser is received for two samples in a row, then a finer quantiser is switched in, or if the largest step is received consistently, then a larger quantiser is used. Both of these techniques are aimed at adapting the method of coding to the data being processed, whereby the modelling of the image is made more spatially localised.

1.4 TEMPORAL PREDICTIVE METHODS

The simple technique of DPCM can be used either in purely the intraframe mode, when it is solely removing spatial redundancy as above, or in interframe mode, or indeed in mixed modes, depending on the definition of the neighbourhood used. In the purely intraframe mode, compression of a ratio of up to about three to one may be achieved with DPCM, whilst ADPCM achieve slightly greater reductions. In the mixed mode, a variation of which is the basis of the majority of coders currently in use in video-conferencing, a larger compression rate may be achieved.

In the mixed inter/intra mode, the choice of neighbourhood is often switched according to which neighbours will give the best prediction. This then requires some side information indicating which neighbourhood is being used. For example, one might switch between pixels in the current frame and pixels in the previous frame. In some implementations the previous frame is only used when the error in prediction is close to zero, and run-length coding of the switching decision can reduce the data-rate drastically. This method is called conditional replenishment and is equivalent to coding only the differences that occur in the temporal domain. This is the technique used for the current generation of video-conferencing equipment, for it is ideally suited to those applications where the camera is fixed relative to the background of the picture, and therefore the area of motion usually represents a small percentage of the picture. As mentioned in section 1.1, this method can code colour pictures at rates of 2 or 1.5 Mbits/s, and indeed can produce reasonable pictures as low as 384 kbits/s, provided there is little movement. This method is the basis of CCITT recommendation H120.

The technique described in this section attempts to model the image sequence in terms of the information content of the sequence, rather than the supposition that the signal is a form of nth-order Markov process, where the information is deemed to occur in those parts of the picture that have changed with time. Although the modelling is of rather a simple type, the method is able to produce a picture of much better quality at a similar data-rate for a particular class of picture.

A more advanced method of prediction which follows on from the concept of coding the temporal differences is a technique called motion compensation. Motion compensations aims to produce a mapping from spatial locations in one frame to spatial locations in a previous frame, thus producing a set of motion vectors which indicate the translation in time of parts of the image. Commonly, blocks of image data are used in practical sytems to allow reasonably robust matching of blocks between images. The efficacy of the matching clearly depends on the content of the image block, for if there is no real structure in the block, then matching is a problem, and if there is structure of a one-dimensional nature, e.g. an edge, then owing to the so-called aperture problem, only the component of the motion across the edge can be found.

Although the concept of motion compensation is philosophically a good one, that is it is effectively tracking the structure within the picture with time, it is too constrained to be used effectively on its own. The main reason for this is that it only models two-dimensional motion in the picture plane, and not the three-dimensional motions that occur in real life. Motion compensation is therefore used with other techniques, which will be discussed next.

1.5 HYBRID METHODS

In coding waveforms, if a high degree of fidelity is attempted, some proportion of the data sent will be used to encode noise. Noise coming from the camera particularly is usually of a high frequency. Texture in an image is also often high frequency, and since the human eye is known to be less sensitive to high spatial frequencies than to low frequencies, the ability to code images in the frequency domain is useful.

This may be achieved on the whole picture or, more usually, on square blocks of size 2^n pixels. Each block is transformed and then coded by a mixture of quantisation of coefficients, only transmitting the coefficients in a certain zone (e.g. ignoring the highest frequency half of the coefficients), and omitting coefficients with low energy. The choice of size of blocks is a trade-off between the localisation of the data and the overhead and numbers of blocks, in that the transform coding of a block is aimed at removing the redundancy within the block: the smaller the block, the greater the proportion of redundancy on average; the more blocks there are, the larger the overall overhead of the picture. In many systems, the choice of blocksize is 8 or 16 pixels.

There are many types of transform that may be used; however, the 'optimal' transform is the Karhunen–Loeve transform, being a method of transforming signals into a set of uncorrelated transform coefficients.

The set of coefficients of the $(N \times N) \times (N \times N)$ Karhunen–Loeve matrix for an $N \times N$ block of pixels is defined by the eigenfunctions of the covariance matrix of the picture array. Thus

$$F(u,v) = \sum_{i=0}^{N=1} \sum_{j=0}^{N=1} F(i,j).A(i,j;u,v)$$

The two-dimensional K–L transform is not in general separable unless the covariance matrix is separable, which is not usually the case. Consequently, because of the high degree of computation, this method is not used extensively. Additionally it can be shown that as N tends to infinity, the Fourier and Discrete Cosine transforms tend to the Karhunen–Loeve transform. As the action of transforming is one of trying to remove spatial correlations in the image, then provided the blocksize is greater than the significant correlations, then these other methods may be used.

In practice, most transforms used are separable, thus allowing a two-dimensional transform to be computed in two orthogonal one-dimensional steps.

The most popular transforms used are those which are most easily implemented, e.g. the Hadamard transform, whose coefficients consist of solely -1s and 1s together with a normalising factor, and cosine and sine transforms, which can be implemented in a fast manner by using butterfly methods, similar to Fast Fourier Transforms.

The cosine transform is particularly favoured because of its approximation to the optimal Karhunen-Loeve transform.

To obtain very low bit-rate video compression systems, where the number of bits per pixel is less than 0.1, say, then usually simply using waveform coding techniques is not enough, for most waveform coding techniques entail sending data for each pixel in the image, giving a data-rate of at least one bit/pixel. Transform coding methods do, however, offer the possibility of sub-unity pixel coding, although it is not easy to use them in the time domain, which is where a great deal of the redundancy occurs. This is due to the desirability of processing two consecutive frames together, rather than 8 or 16, which would entail a great deal of delay in processing, which is unacceptable when accompanied by sound. In addition, if the time sequence is segmented into blocks, then one runs the risk of a mismatch between time segments, due to approximations, which would be objectionable to the viewer.

Therefore a mixture of prediction in the temporal domain and transformation in the spatial domain is often favoured, with motion compensation being a suitable predictive technique to use. The simplest way to implement the motion compensation is to do a block-match of the data. This is also suitable to use because of its compatibility with the subsequent block transform. Thus, the image is first split into blocks; then each block is examined to see if it is different from the same block in the previous frame: if so, then nothing is transmitted. Those blocks that are different are then motion compensated, and their motion vector is transmitted. The prediction is done on that basis, and finally each block is examined to determine the error function, which, if significant, is transformed and the coefficients are quantised and coded as efficiently as possible.

In this video coding approach, the modelling of the information content in the temporal domain is better than the motion compensation method described previously, in that the motion of objects in the image is modelled, albeit in a block-matching way. The errors produced by the motion compensation are then transformed and coded, perhaps representing less a feat of modelling than a pragmatic method of achieving a low bit-rate for low energy blocks.

1.6 CONCLUSIONS

In video coding to date, great progress has been made in the last five years, with the possibility of large compression rates with reasonable quality coming closer and closer. Further compression in this area will be possible only if images and their contents are able to be modelled more accurately and compactly, particularly with respect to those features that the eye finds important.

With a coding strategy increasingly geared to the structure within an image, the data-rate for transmission is extremely variable; this implies a control strategy for allocation of data bits to those parts of the image which are most perceptually important, giving rise to an important part of coding research.

The advent of very fast integrated circuits, performing complicated functions such as cosine transforms at video rates, combined with developments like the transputer, make more complicated modelling increasingly possible.

In the long term, these elements must give rise to a greater 'video society'.

REFERENCES

[1] C. E. Shannon & W. Weaver, *The Mathematical Theory of Communication*, (1949) University of Illinois Press.

2

Speech and image signals: recognition as classification under uncertainty

M. Ingleby
Department of Computer Studies and Mathematics, The Polytechnic, Huddersfield.

2.1 MATHEMATICAL TAXONOMY

A mathematical taxonomy [1] (Deterministic) is often also called *numerical taxonomy* [2], *automatic taxonomy* [3] on *cluster analysis* [4]. This is the study of ways of classifying a population of individuals according to their resemblance in certain of their properties. A taxonomic study usually begins with a table of the form shown in Fig. 2.1, in which C_{ij} is the measure of property P_j for individual x_i.

Individual ↓	Property → P_1	$P_2 \ldots$	$P_j \ldots$	P_M
x_1	C_{11}	$C_{12} \ldots$	$C_{1j} \ldots$	C_{1M}
x_2	C_{21}	$C_{22} \ldots$	$C_{2j} \ldots$	C_{2M}
$\vdots$	$\vdots$	$\vdots$	$\vdots$	$\vdots$
x_i	C_{i1}	$C_{i2} \ldots$	$C_{ij} \ldots$	C_{2M}
$\vdots$	$\vdots$	$\vdots$	$\vdots$	$\vdots$
x_N	C_{N1}	$C_{N2} \ldots$	$C_{Nj} \ldots$	C_{NM}

Fig. 2.1 — Taxonomy table.

Individuals are represented in the table by rows of numbers, elements of a normed space of dimension M, and one can introduce various distance functions on this space which tell the extent to which two individuals differ:

Example 1: $d_1(x_i,x_k) = \text{Max}\ \{|C_{ij} - C_{kj}| : j = 1,2, \ldots, M\}$

Example 2: $d_2(x_i,x_k) = \sqrt{\Sigma_j(C_{ij} - C_{kj})^2}$

From the distance function one can obtain coefficients of resemblance: if $d(x_i,x_j)$ has a greatest value D, we can define a coefficient of resemblance, for example, as

$$r_1(x_i,x_j) = \frac{D - d(x_i,x_j)}{D + d(x_i,x_y)}$$

with the property $0 < r_1(x_i,x_j) < 1$, the upper bound being achieved for identical individuals, the lower bound by the pair of individuals differing most from each other.

Coefficients of resemblance can be defined directly, too — for example

$$r_2(x_i,x_k) = \tfrac{1}{2}\left[\frac{(\Sigma_j C_{ij}C_{kj}) + 1}{\sqrt{(\Sigma_j C_{ij}^2)\ (\Sigma_j C_{kj}^2)}}\right]$$

has the property of $0 < r_2(x_i,x_j) < 1$.

A resemblance coefficient can be processed into a *likeness relation* by introducing a likeness *threshold* α:

$$x_i \text{ is like } x_j \text{ iff } r(x_i,x_i) > \alpha$$

Such a notion of likeness constitutes a reflexive, symmetric binary relation in the set of individuals (images, sounds, etc.). In order to partition a set, a binary relation must be an equivalence relation, that is it must be reflexive, symmetric and *transitive*. The various commercial/scientific software packages contain different ways of forcing transitivity in a not-necessarily-transitive relation (and are often inefficient through lack of knowledge/use of the algebra of relations in their specification or implementation).

In the algebra of relations [5], it is known that every relation has a minimal (i.e. weakest) transitive extension which can be constructed iteratively. The construction is easily summarised using representative Boolean matrices as follows:

> *Let R_{ij} = TRUE if $r(x_i,x_j) > \alpha$, else R_{ij} = FALSE*
> *Let $T = S = R$*
> (1) *Let $T = S \circ S$*
> *If $T = S$ then T is the required extension*
> else *Let $S = T$ and return to* (1).

Recursive formulations are possible too, and in some computer languages are more efficient.

With threshold α set at 0, all individuals are like, and the whole population falls into a single class. With threshold α set at 1, all pairs of distinct individuals are unlike, and each individual is in its own class, the only member of its class. As α decreases slowly from 1, individuals associate in classes, then classes associate, and so on; the whole process of association being summarised in a *dendrogram* as shown in Fig. 2.2.

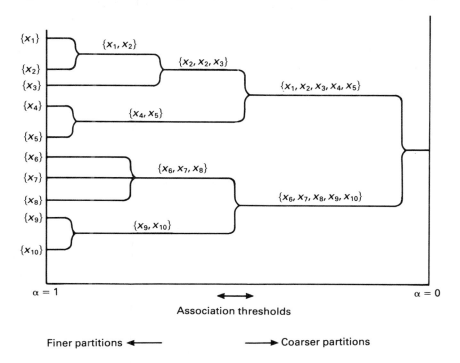

Fig. 2.2 — Example of a dendrogram.

A dendrogram is a graphical realisation of a more abstract mathematical entity which encapsulates the essence of classification. This entity is a *chain* of ever-coarser partitions [6], and the chaining takes place in the family of all partitions of the set of individuals being classified. Such a family, partially ordered by the relation 'coarser than', has the mathematical structure of a *lattice*. Lattices of partitions (or, isomorphically, lattices of equivalence relations in a set) were thoroughly examined by the distinguished algebraist Oystein Ore (of four-colour problem renown) in the 1940s [7]. But before looking at Ore's results and their consequences, a glance at the property tables relevant to speech and image classification is timely.

2.2   PROPERTY TABLES FOR SPEECH AND IMAGE CLASSIFICATIONS

The previous chapter by Mason ended by indicating a future need to consider the content of a visual scene in order to communicate it to distant humans, with increased technical efficiency. Traditional methods of communication have not interpreted scenes for content, or sound waves for speech content. They have transduced a signal and facsimiled it for subsequent interpretation by a human end-user. As soon as one begins to automatically analyse scenes and sounds for content, whether with bettter communication in view or increased autonomy for seeing/hearing robots, certain perplexing features of the recognition process itself present themselves. This chapter describes these features, and locates them in the process of taxonomic description.

To begin with speech, the individuals in a property table for speech classification will be *phonemes*. These are fragments of speech separated by short pauses. Continuous speech can be understood as phoneme strings. Phonemes are the focus of an agreement between structural phoneticists — who name them by the organs of speech used to make them (teeth, lips, palate, tongue, . . .) — and acoustic phoneticists — who characterise them by frequency–time–amplitude pictures — (dealt with more fully in the next chapter).

The properties of a speech fragment which characterise it as a phoneme include:

energy parameters	— total pressure fluctuation energy, energy in a broad-spectrum '*unvoiced*' part of the pressure signal, . . .
frequency parameters	— mean frequency over phoneme duration of two or three *formants* of '*voiced*' parts of the pressure signal (formants are resonances of the speaker's vocal tract), mean frequency and variance of the unvoiced part of the signal, . . .
elision parameters	— rise-time of new formants from formants of the previous phoneme, fall-time towards formants of next phoneme, transients, . . .
suprasegmental parameters —	relating to patterns of intonation and stress in a string of phonemes.

A classifier of speech will have to be very circumspect in choosing a distance function. It must, for example, be defined in terms of *relative* parameters of energy and frequency so that intentionally similar utterances differing in pitch, timbre and loudness will be metrically close (and hence taxonomically like). On the other hand, it must metrically separate signals which are very like in physical properties but different in intention: 'Lettuce' versus 'Let us', . . . 'offal' versus '. . . awful' versus '. . . of all' . . . . These opposed requirements will involve the classifier in, as we shall see, both ambiguity of interpretation and paraodixicality.

When classifying images, the choice of metric/resemblance coefficient is no less problematic. Individuals will be scenes captured by a video camera, for example. In a simplistic view, properties might be the contents of pixel locations representing a scene, or perhaps coarse-grained averages of these. If, however, an object changes location in a scene because of relative motion with respect to the data capture device, the pixel map changes radically and metrics must take such changes into account (i.e. by according them little interpretative significance). The current art of image interpretation tends to achieve this by low-level (possibly parallel-) pre-processing of pixel contents to synthesise properties related to the visual primitives of human or mammalian psychophysics. Humans are, after all, supremely successful classifiers of scenes via the objects or *primitives* they 'see' in scenes, the 'seeing' going on at the back of the head (visual cortex) as much as immediately behind the retina [8].

Examples of automatic low-level processors are

edge detectors — which construct lines following luminance edges in a scene, thereby coding the outlines of objects, . . .

scaling — which accounts for the distance-with-optical-size variation of objects and is based on geometrical optics,

luminance valley detectors — which produce lines following, for example, the wrinkles in a video picture of a face, thereby nicely complementing pure silhouette information coded by an edge detector,

other detectors — texture gradients, colour contrast peaks and valleys, etc.

Once again the difficulty of defining metrics to sort these out to discriminate where appropriate and ignore differences which are not interpretationally significant leads one into ambiguity and paradox, to which, as we shall see, our own psychophysical classifiers are not immune.

2.3 PROPERTIES OF PARTITION LATTICES

The lattice of partions of the set $\{1,2,3\}$ has only four members:

$$1 = [\{1,2,3\}]; \quad A_1 = [\{1\}, \{2,3\}]; \quad A_2 = [\{2\}, \{1,3\}];$$
$$A_3 = [\{3\}, \{1,2\}]; \quad 0 = [\{1\}, \{2\}, \{3\}]$$

The lattice operations are denoted $X \vee Y$, for the finest partition coarser than both X and Y, and $X \wedge Y$, for the coarsest partition finer than both X and Y; and both operations can be described in the same table, exploiting their commutativity in the way used, as, for example, in [9].

$\vee$ \ $\wedge$	0	A_1	A_2	A_3	1
0	0	0	0	0	0
A_1	A_1	A_1	0	0	A_1
A_2	A_2	1	A_2	0	A_3
A_3	A_3	1	1	A_3	A_3
1	1	1	1	1	1

Lattices are often described using a Hasse diagram which, for the above partition lattice, takes the form shown in Fig. 2.3. This diagram serves to illustrate the following terms taken from the theory of lattices:

atoms are elements minimally coarser than the finest element 0—in the diagram, A_1, A_2 and A_3 are atoms;

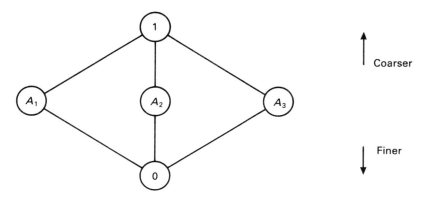

Fig. 2.3 — Hasse diagram for partition lattice of the set {1,2,3}.

an atomic lattice is a lattice of all whose non-zero elements are joins (∨) of the atoms which they contain;

an independent set of atoms is a non-empty set for which no member lies in the join of other members (cf. linear independence in vector spaces, where no vector from an independent set lies in the span of other members);

a *basis* is an independent set of atoms which is maximally independent in the sense that the addition of one further atom to the set destroys independence in the larger set; for example, sets $\{A_1\}$ and $\{A_1,A_2\}$ are independent in the above lattice, but only the latter is a basis because adding a further atom, A_3, produces $\{A_1,A_2,A_3\}$ which is dependent because A_3 lies below $A_1 \vee A_3 = 1$ in the Hasse diagram.

It turns out that all partition lattices are atomic and have finite bases, the latter serving as a kind of dimension theory in such lattices: the *dimension of a lattice* being the size of any one of its bases, the *dimension of an element* in such a lattice being the size of the greatest independent set of atoms lying within the element. The Hasse diagram of the partition lattice of the set {1,2,3,4} allows one to verify the sense and consistency of these notions.

   The partitions are labelled thus:

minimal element, $0 = [\{1\}, \{2\}, \{3\}, \{4\}]$;
atoms (of dimension 1), $A_{12} = [\{1,2\}, \{3\}, \{4\}]$,
$A_{34} = [\{3,4\}, \{1\}, \{2\}]$, $A_{14} = [\{1,4\}, \{2\}, \{3\}]$,
$A_{23} = [\{2,3\}, \{1\}, \{4\}]$, $A_{24} = [\{2,4\}, \{1\}, \{3\}]$,
$A_{13} = [\{1,3\}, \{2\}, \{4\}]$;

co-atoms (of dimension 2), $C_1 = [\{1\}, \{2,3,4\}$, $C_2 = [\{2\}, \{1,3,4\}]$
$C_3 = [\{3\}, \{1,2,4\}]$, $C_4 = [\{4\}, \{1,2,3\}]$, $B_{12} = [\{1,2\}, \{3,4\}]$,
$B_{13} = [\{1,3\}, \{2,4\}]$, $B_{14} = [\{1,4\}, \{2,3\}]$;

maximal element, $1 = [\{1,2,3,4\}]$

and the Hasse diagram is given in Fig. 2.4.

The latice is three-dimensional by virtue of the fact that $\{A_{12}, A_{34}, A_{14}\}$ forms a basis. It also has an interesting sublattice, with Hasse diagram given in Fig. 2.5 — a sublattice which also occurs in the partition lattice of all the sets with cardinal number > 3. The existence of such sublattices is a guarantee that the *distributive law* $X \wedge (Y \vee Z) = (X \wedge Y) \vee (X \wedge Z)$ *fails*, as can be seen from the counter-example:

$$\left. \begin{array}{l} C_1 \wedge (B_{14} \vee C_4) = C_1 \wedge 1 = C_1 \\ C_1 \wedge B_{14} = A_{14}, \ (C_1 \wedge B_{14}) \vee (C_1 \wedge C_4) = A_{14} \\ C_1 \wedge C_4 = A_{14} \end{array} \right\} \quad \text{unequal}$$

The failure of this important law is reminiscent of a similar failure in *geometric lattices* — a geometric lattice being the lattice of subspaces of a finite-dimensional vector space over a field.

Fig. 2.6 shows the situation for $\mathscr{L}(\mathbb{R}^2)$, the lattice of subspaces of the two-dimensional real vector space $\mathbb{R}^2$. The atoms are lines through the origin and include the axes X and Y and lines $Z(\theta)$ inclined at the angle θ. The Hasse diagram is given in Fig. 2.7. with continuously many non-distributive sublattices. If the field $\mathbb{R}$ is replaced by one of the finite fields, for example those used in coding theory, the number of atoms becomes finite.

The work of Ore [7] and successors [9] shows that the lattices $\mathbb{P}_n$ of partitions of sets $\{1,2, \ldots ,n\}$ are isomorphic to geometric lattices. Indeed, an isomorphism ϕ: $\mathbb{P}_3 \to \mathscr{L}(\mathbb{Z}_2^2)$ is easy to contrive:

$$\begin{array}{l} \varphi(0) = \{(0,0)\}, \quad \varphi(1) = \{(a,b) : a,b \in \mathbb{Z}_2\}, \\ \varphi(A_1) = \{(a,0) : a \in \mathbb{Z}_2\}, \quad \varphi(A_2) = \{(0,a) : a \in \mathbb{Z}_2\}, \\ \varphi(A_3) = \{(a,a) : \ \in \mathbb{Z}_2\} \end{array}$$

where $\mathbb{Z}_2$ is the field with only two elements. Table 2.1 describes another isomorphism $\psi : \mathbb{P}_4 \to \mathscr{L}(\mathbb{Z}_2^3)$.

The notion of *mutual compatibility* mentioned in Table 2.1 is a lattice theoretic correlative of the concept of mutually compatible observables in quantum physics. Two elements in a lattice are by definition mutually compatible if there is an ascending chain in the lattice containing both those elements. An important feature of partition lattices is the occurrence of mutually incompatible partitions — pairs of partitions which cannot lie in the same chain, or in the language of mathematical taxonomy, pairs of partitions which can in no circumstances arise in the same dendrogram. Such occurrences have profound consequences when one wishes to describe uncertainty and ambiguity in taxonomic studies, and, quite fortunately, there is an historical circumstance of similar occurrences which have been successfully and practically dealt with by one of the greatest mathematical intellects of this century, von Neumann [10].

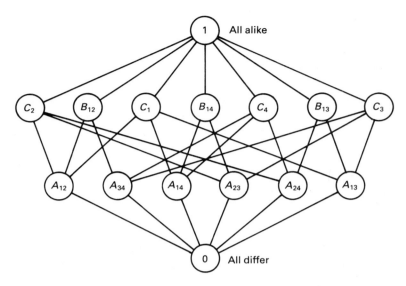

Fig. 2.4 — Hasse diagram for lattice of set $\{1,2,3,4\}$.

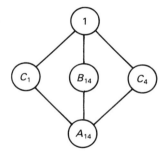

Fig. 2.5.

2.4 AN HISTORICAL NOTE

Two complex 'wavefunctions' of a quantum mechanical system describe the same state if one is a complex multiple of the other; as 'states' they are atoms — one-dimensional subspaces — in the lattice of closed subspaces of a complex Hilbert space [1]. Many years ago, Birkhoff and von Neumann [12] showed that the above state-space lattices or 'qauntum logics' are non-Boolean in the sense of failure of the distributive law. Subsequently von Neumann went on to develop appropriate statistics on such lattices — a way of describing uncertainty into which Heisenberg's uncertainty principle was incorporated, at the expense of a radical departure from

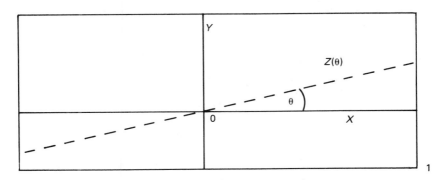

Fig. 2.6 — The lattice of subspaces of $\mathbb{R}^2$.

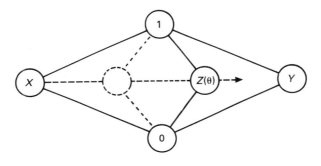

Fig. 2.7.

the conventional probability theory based on Kolmogorov's [13] concepts of 'random model' and 'logic of events'.

Von Neumann uncertainty descriptors, called 'density matrices', are no longer radical or 'theoretical' entities; the probabilities they encompass are related to real frequencies of real occurrences measured with laboratory instruments interfaced to very practical microphysical systems. They remain, however, qualitatively different from the uncertainty descriptors *à la Kolmogorov*, the differences having been pointed out by Mackey [14] and exploited by the present author elsewhere [15]. In Table 2.2, these differences are summarised in terms which deliberately recall the ideas of Schafer and Dempster [16] regarding the *theory of belief* and the way belief is altered by the onslaught of new evidence on the (rational) believer.

The isomorphisms of the previous paragraph allow one to map von Neumann uncertainty descriptors from a geometric lattice to a partition lattice, and the probabilities obtained from such a mapping are tabulated alongside the isomorphism in section 2.3. They were obtained from a density matrix on $\mathscr{L}(\mathbb{Z}_2^3)$ assigning probability P_1 to the ray $X = \{(a,0,0):a \in \mathbb{Z}_2\}$ and probabilities P_2 and P_3 respecti-

Table 2.1 — Probabilities assigned to the lattice $\mathbb{P}_4$. The probabilities are those for a density matrix with diagonal elements $P_1, {}_{P2}P_3$ in basis i, j, k of $\mathbb{Z}_2^3$

Elt. of $\mathbb{P}_4$	Isomorphic image in $\mathscr{L}(\mathbb{Z}_2^3)$	nature of element	Probability
A_{12}	$\{(a,0,0):a\in\mathbb{Z}_2\}$	Atoms in a	P_1
A_{23}	$\{(0,a,0):a\in\mathbb{Z}_2\}$	given basis, all	P_2
A_{34}	$\{(0,0,a):a\in\mathbb{Z}_2\}$	mutually compatible	P_3
A_{13}	$\{(0,a,a):a\in\mathbb{Z}_2\}$	Atoms incompatible	Unassigned
A_{14}	$\{(a,a,a):a\in\mathbb{Z}_2\}$	with basis atoms	Unassigned
A_{24}	$\{(a,0,a):a\in\mathbb{Z}_2\}$		Unassigned
B_{12}	$\{(a,b,0):a,b\in\mathbb{Z}_2\}$	Joins of two	$P_1 + P_2$
C_2	$\{(a,0,b):a,b\in\mathbb{Z}_2\}$	compatible basic	$P_1 + P_3$
C_3	$\{(0,a,b):a,b\in\mathbb{Z}_2\}$	atoms	$P_2 + P_3$
B_{13}	$\{(a,b,a+b):a,b\in\mathbb{Z}_2\}$	Joins of	Unassigned
B_{14}	$\{(b,b,a):a,b\in\mathbb{Z}_2\}$	incompatible	Unassigned
C_1	$\{(a,b,b):a,b\in\mathbb{Z}_2\}$	atoms	Unassigned
C_4	$\{(b,a,b):a,b\in\mathbb{Z}_2\}$		Unassigned
0	$\{(0,0,0)\}$	Smallest element	?
1	$\{(a,b,c):a,b,c\in\mathbb{Z}_2\}$	Largest element	$P_1 + P_2 + P_3$

Table 2.2

Feature	Kolmogorov	von Neumann
'Frame of discernment'	Set **S** of possible outcomes . . .	Complex Hilbert space **H** of state vectors
Elementary Events	Points of set **S**	Atoms in the lattice $\mathscr{L}(\mathbf{H})$
Compound events	Subsets of **S**, these forming a Boolean σ-algebra $\mathscr{B}(\mathbf{S})$	Closed subsets of **H**, these forming a quantum logic $\mathscr{L}(\mathbf{H})$
Belief function or 'uncertainty descriptor'	Probability measure on $\mathscr{B}(\mathbf{S})$ — a σ-additive function with prob $(\mathbf{S}) = 1$	A density matrix or operator on **H**, diagonal with respect to an orthogonal basis

vely to $Y = \{(0,a,0):a \in \mathbb{Z}_2\}$ and $Z = \{(0,0,a):a \in \mathbb{Z}_2\}$. No probability is assigned to those lattice elements incompatible with these mutually compatible atoms — or in taxonomic terms, those partitions which could never occur in dendrograms containing the partitions $\psi^{-1}(X), \psi^{-1}(Y), \psi^{-1}(Z)$. This is the kind of uncertainty descriptor one needs to describe uncertainty in the clustering of images for likeness, phoneme strings for equivalences of semantic intention, and so on.

2.5 AMBIGUITY, PARADOX AND LOGIC OF RECOGNITION

Having presented recognition as a taxonomic event, and noted that the lattices underlying taxonomy are non-distributive, one must be prepared to find non-distributive features in the recognition process, visual and auditory. In order to appreciate such features, some well-known features of the psychophysics of perception must be recalled. In the area of vision, these features have been described by Gombrich [17] and others from whose work these figures are taken.

Image 1 represents a transparent cube with coloured edges. The upper and lower

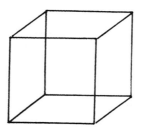

Image 1
Ambiguous images:
Which face is foremost?

square faces could both be understood as the front face, and our human recognition system tends to oscillate in its preferred interpretation rather as if the line drawing had driven it into a limit cycle state. Image 2 is the duck–rabbit, and the geometrical Image 3 is either a rectangular object illuminated from above to generate shade, or a skewed object with parts painted darker and lighter grey.

The *paradoxical* images 4 and 5 present no problems as sets of lines, but when our recognisers set to work on them there is conflicting of data from one part of the image to another — the pictures seem to twist and turn as we direct attention to different parts.

Similar things can happen with phoneme strings:

Ambiguity:
1. 'LET US PRAY AND ALL LIVE SINLESS, BETTER . . . '
   ‹‹Lettuce spray and olives in less bitter . . . ››

Image 2
Rabit or duck?

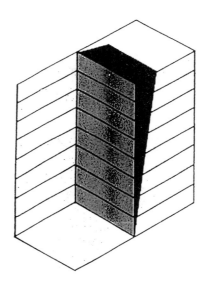

Image 3
Rectangular or skewed?

Ambiguity:
2. 'PASS IAN'S SILVER PLATE!'
   ‹‹patience, s'il vous plaît››

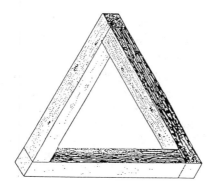

Image 4

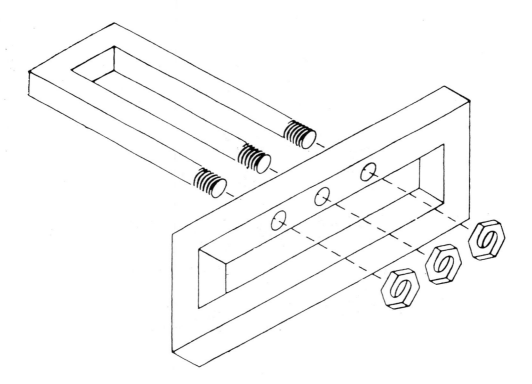

Similar things can happen with phoneme strings:

Image 5

Paradox:

3. *'Twas brillig and the slithy toves*
   Did gyre and gimbal in the wave'. (*From Jabberwocky* by Charles Lutwidge
   Dodgson (Lewis Carroll).)

The two strings in the first example would be extraordinarily difficult to distinguish using an automatic speech recogniser because the physical similarity of sounds is very great in spite of dissimiliarity of semantic intention. The second example draws attention to the fact that recognition/classification depends very much on the set of primitives in the head, different for different cultural contexts. There is a famous [17] visual example of culture-dependent seeing, given in Image 6.

Image 6
Family Group — window or oil-container added?

Most Europeans interpret the scene in Image 6 as a family group indoors, with a window in the background. Many Africans, however, have interpreted the scene as a family group *outdoors*, the eldest girl carrying an oil-container or her head!

In the third example, the *Jabberwocky* phoneme string produces sensations of paradox, which have been used to humorous effect — for example by Stanley Unwin — by offering to our auditory systems phoneme strings of near-words for which there is no consistent interpretation. For example, 'slithy tove' could be a deformed version of 'lithe dove' or 'slimy toad', both of which might conceivably gyrate, but neither of these would be seen to 'gimbal in the wabe' if by this one meant 'gambol in the glade' or 'gambol in the wave' or, as Lewis Carroll hints, 'gimblet in the glebe'.

Returning, after these digressions, to distributive laws and the logic of recognition, it is possible to exploit ambiguity and paradox in vision to obtain counter-examples showing that the distributive law fails. One such is concerned with pictures of dice, and it is necessary to appreciate that dice with faces numbered so that opposite faces always bear a total of seven points (six opposite one, four opposite three, etc.) have *parity*. If the dice in Fig. 2.8 are assembled, the left-side die can be

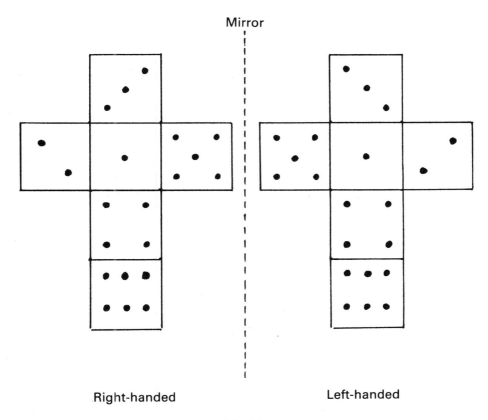

Fig. 2.8.

placed beside the left hand, thumb directed perpendicular to faces 1 and 6 from 1 to 6, fingers curling from 2 to 2. The right-hand die, however, requires a right hand beside it to have thumb directed 1 to 6 and fingers curling 2 to 3. Three-dimensional objects may or may not have parity, as physicists well know. Indeed, most of us know that if we are asked to obtain a left-handed pair of scissors, this is by no means impossible, but it we are sent for a left-handed spanner or screwdriver, then someone is having a joke at our expense. Some readers who are used to image communication may also know that specular reflection changes the parity of an object which, unlike the screwdriver, is not parity neutral.

The counter-example the author wishes to present is concerned with pictures A, B, C of dice given in Fig. 2.9. We can combine these images using some (binary)

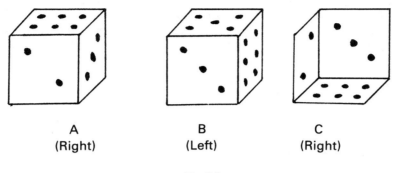

A	B	C
(Right)	(Left)	(Right)

Fig. 2.9.

lattice operations which the author has been exploring elsewhere in work on modelling under uncertainty. The image $A \vee B$, the join of two images A and B, is the most certain image less certain than both A and B. Graphically it is made up of the parts that A and B have in common. Using the images of dice above,

$A \vee B$ is

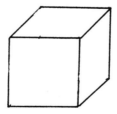

which is ambiguous and parity neutral.

The other lattice operation is defined such that $A \wedge B$, the meet of two images A and B, is the least certain image more certain than both A and B. Graphically it is produced by superposition. The superposition of images which contradict — for example, by having opposite parity — can produce a paradoxical image which in this lattice based on the relation 'more certain than' would be symbolised by 0. An empty visual field would be symbolised by 1.

Fig 2.10 and Table 2.3 summarise the results of forming the lattice compounds $(A \vee B) \wedge C$ and $(A \wedge C) \vee (B \wedge C)$ — these should be equally certain if the distributive law is to hold.

In Table 2.3 a '—' signifies that the number of spots on the face in question is

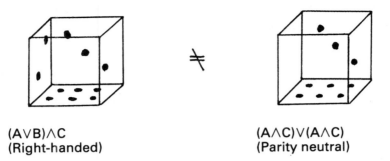

(A∨B)∧C
(Right-handed)

(A∧C)∨(A∧C)
(Parity neutral)

Fig. 2.10.

Table 2.3

Image	$A \lor B$	$(A \lor B) \land C$	$A \land C$	$B \land C$	$(A \land C) \lor (B \land C)$
Spots on:					
Top face	—	(1)	6	5	(1)
Bottom face	—	6	6	6	6
Right face	—	(5)	2	6	—
Left face	—	2	3	2	—
Front (if upper square foremost)	×	3/ ×	3/ ×	3/ ×	3/ ×
Front (if lower square foremost)	—	× /(4)	+ /2	× /3	× /(4)
Remarks	Lower square foremost	Ambiguous image	Ambiguous geometry, paradox in spot numbers	Ambiguous geometry, paradox in spot numbers	Ambiguous geometry consistent spot numbers

uncertain (i.e. the face is *obscured*). Also, a '×' signifies that a face whose relative position is ambiguous (front or back) is to be interpreted in the opposite way to that stated in the left-most column of the table. Finally a number in parentheses (e.g. (4) or (1)) is inferred from what is seen on the opposite face rather than being directly seen on the image.

As the images in Fig. 2.10 emphasise, the images $(A \lor B) \land C$ and $(A \land C) \lor (B \land C)$ are unequal, the former image being strictly more certain than the latter. There rests the case that, because recognition is classification and partition lattices are non-distributive, the algebra of recognition events is a non-Boolean algebra and the description of uncertainties of recognition must be of the von Neumann type rather than the Kolmogorov type.

ACKNOWLEDGEMENTS

The discussion of the final counter-example has been sharpened and amplified in editing for publication, in response to very helpful and constructive comments by a number of conference participants.

REFERENCES

[1] N. Jardine & R. Sibson, *Mathematical Taxonomy*, (1971) Wiley, London and New York.
[2] R. R. Sokal & P. H. A. Sneath, *Principles of Numerical Taxonomy*, (1963) Freeman, London.
[3] M. Jambu & M. O. Lebeaux, *Classification Automatique pour l'Analyse des Donnees*, (1984) Dunod, Paris.
[4] B. Everitt, *Cluster Analysis*, (1974) Heinemann/SSRC, London.
[5] G. Gratzer, *Universal Algebra*, (1968) van Nostrand, New York.
[6] O. Ore, 'Chains in partially ordered sets', *Bull. Amer. Math. Soc.*, **49**, 558–556 (1943).
[7] O. Ore, 'Theory of equivalence relations', *Duke, Math. J.*, **9**, 573–627 (1942).
[8] D. Marr, *Vision*, (1982) Freeman, New York.
[9] G. Grätzer, *General Lattice Theory*, (1978) Academic Press, New York.
[10] J. von Neumann, *Mathematical Foundations of Quantum Mechanics*, (1955) Princeton UP, New Jersey. *Thermodynamik quantenmechanischer Gesamthei-ten*, Gottinger Nachricten, pp. 273–291, (1927).
[11] V. S. Varadarajan, *The Geometry of Quantum Theory*, (1970) (2 vols), van Nostrand, New Jersey.
[12] G. Birkhoff & J. von Neumann, 'The logic of quantum mechanics, *Ann. Math*, **37**, 823–843 (1936).
[13] A. N. Kolmogorov, *Foundations of the Theory of Probability*, (1950) Chelsea, New York (translation of earlier German text).
[14] G. W. Mackey, *The Mathematical Foundations of Quantum Mechanics*, (1963) Benjamin, New York.
[15] M. Ingleby, 'Some criticisms of quantum logic', *Helv. Phys. Acta*, **44**, 299–314 (1971). 'Non-Boolean event algebras in the probabilistic description of classical boundary and initial-value problems', in *Modelling under Uncertainty* 1986 (eds Jones and Davies), Institute of Physics, (1986).
[16] G. Schafer & A. Dempster, *The Mathematical Theory of Evidence*, (1982) Princeton University Press.
[17] E. Gombrich, *Art and Illusion: Study in the Psychology of Pictorial Represen-*

tation, (1977) Phaidon, Lon. *The Image and the Eye: Further Studies in the Psychology of Pictorial Representation*, (1982) Phaidon, Oxford, (with R. L. Gregory, joint editor) *Illusion in Nature and Art*, (1973) Duckworth, London.

3

Identification of speech in noise using low-cost algorithms

W. H. Godwin, M. J. Herring, D. P. Goodall and **W. Bates**
Department of Computing, Electronics and Mathematical Studies,
Gloucestershire College of Art and Technology, 73 The Park, Cheltenham, Glos.
GL50 2RR

3.1 INTRODUCTION

Modelling of speech signals has received a great amount of attention from the perspective of the joint problems of automatic Speech Recognition and Speech Synthesis.

The difficulty of the problem has led to the concentration on the use of fast computers exploiting parallelism in its various forms. Rather less attention has been given to low-cost computing implementation, since microprocessors in personal computers have generally not been capable of robust recognition.

Identification of Speech in Noise is a problem with a different perspective: where automatic recognition is impractical, the computer can be used to identify those segments of a signal which are likely to contain speech, and provide some measure of intelligibility. The model must deal with connected speech by an unknown speaker in an unknown language in an unknown acoustic environment subject to interference by a variety of unpredictable noise sources, from steady white noise to time-varying quasi-periodic noise.

A low-cost approach (implemented on a PC-compatible micro) forces careful evaluation of the necessity of any calculation performed. An optimisation problem is involved in which the cost of a calculation is balanced against the extra information which it is likely to provide.

The high uncertainty in the model inevitably limits the accuracy of estimation. An important criterion is that the system should not fail to spot an isolated syllable; but some mislabelling of noise events as speech is acceptable.

Section 3.2 describes the techniques used to keep the computational cost low.

Section 3.3 describes the modelling of the signal. Section 3.4 describes computational methods of feature analysis, and section 3.5 discusses problems arising.

3.2 LOW-COST TECHNIQUES: HIERARCHY

3.2.1 General perspective on organisation for cost optimisation

Given a segment of the signal which might contain speech, the problem is one of organising an efficient search for intelligibility; this search occurs in the time dimension and in feature dimensions for various local transformations of the signal. Principally the search is for features more likely to be present in speech than in noise.

The system proposed includes a set of Feature Extraction Algorithms. The application of these algorithms is subject to dependencies:

(1) High/low. Higher algorithms are calculated using the results of lower algorithms.
(2) Fine/coarse. Finer features are present in better quality speech signals.
(3) Costly/cheap. Costlier algorithms consume more processor time.

Out of these three dimensions an approximate hierarchy can be arrived at which reduces the tendencies to:

(a) repeat a calculation in a different algorithm;
(b) look for a feature in an unlikely time slice;
(c) employ complex means to extract small amounts of information.

3.2.2 Cost reduction by selective application

The features are extracted in sequence according to the hierarchy, each algorithm being applied only to time slices which previous analyses have shown to be likely to contain speech. Further selectivity, in the time dimension, results from consideration of expected time distribution of speech events (syllable durations, silent intervals). This is done in an *ad hoc* manner, though there are avenues for systematisation [1], subject to experimentation.

3.2.3 Cost reduction by restriction of accuracy

Quantisation of the digitally encoded signal leads to the use of look-up tables and simple integer arithmetic in preference to floating-point implementation. The number of levels of quantisation can sometimes be beneficially reduced to keep calculation within 16-bit capability without serious degradation in estimates.

3.2.4 Cost reduction by partial calculation

Certain iterative calculations can be reduced if there is *a priori* information as to the expected value of the feature. The reduction is effectively in the size of the search space. This arises as a result of the continuity largely present in specch signals. There is a loss in accuracy of the feature computed.

3.3 MODELLING OF SPEECH, SIGNALS AND DISCRIMINATION

The signal itself is, in our case, a time series of sample values in the range [− 3096, + 3096], logarithmically compacted (quantised) into an 8-bit code, with a sampling rate of 8000 Hz.

3.3.1 Model of speech

One approach is to model the production of speech, varying the model's parameters to optimise and fit between the model's output and the speech waveform. The other approach is to model perception by simulating the transformations carried out by the ear on incident sound waves. The latter is judged more influential to the problem in hand, though the two approaches are sometimes complementary.

3.3.2 Perception

The ear responds to variations in air pressure [2]. Repeated events up to 100 Hz cause firing of nerve cells at the frequency rate, but above this point the frequency is primarily coded by location of the firing cells. Hence there are two modes for pitch perception. The range 600–1500 Hz corresponds to the region on the ear's membrane with largest displacement. Amplitude follows an approximate logarithmic perceptual scale. For spectral perception there is a short-term windowing effect. The ear responds rapidly to transience of the signal, but takes several ms to adapt and perceive the new total structure [3]. The separation and interaction of time domain and frequency domain effects is an important factor in the modelling problem.

3.3.3 Production

The anatomy of the vocal tract enables the production of a pressure wave containing a time-varying spectrum which can carry information at a rate up to 50 bits/s [1].

   The vocal chords, energised by air from the lungs, produce an excitation signal, often periodic and pulse-like in nature, which is coloured by the shape of the vocal tract. This approximates to a series of tubes, varying in diameter, with side branches. Lips, tongue, velum, pharynx and vocal chords are used to restrict the air flow and provide articulation of the sounds, for stop and friction effects. The resulting sound can be treated as a sum of exponentially damped sinusoids.

3.3.4 Transmission and recording

Electrical transmission of the signal leads to deterioration of its characteristics, under the general heading of noise. This comprises:

(a) Added extraneous events.
(b) Change in spectral envelope: loss of low and high frequencies.
(c) Distortion: modulation of some frequencies by others.

3.3.5 Speech discrimination model

Intelligibility depends on the ear's ability to classify events into different phonetic categories. Analysis of speech reveals characteristics (features) of the waveform which serve as discriminants for phonetic classification.

ENERGY   — Short-term rms value of the signal.
VOICING   — Quasi-periodic events, below 1 kHz.
FRICTION   — Non-periodic noise at higher frequencies.
FORMANTS   — Spectral prominences above 200 Hz.
VARIATION of features especially:

PITCH variation of voice,
FORMANTS — mainly the three lowest,
ONSET TIMING of voicing.

Some voicing (vibration of vocal cords) is expected in all syllables unless the speech is generally unvoiced (whispered). The pitch (F0) is usually above 50 Hz. Formant distribution characterises vowels. The first formant (F1) normally falls in the range 200–800 Hz, the second formant (F2) at 600–2400 Hz, F3 at 2000–3000 Hz, with bandwidth som 50 Hz. Friction and onset time are important to some consonants. Formant variation is important in all consonants.

3.3.6   Observation of features
It is convenient to observe short-term features in terms of frames: a frame is a consecutive set of samples covering a number of milliseconds. Frame length may be varied, and frames may overlap.

Energy is a short-term average amplitude; voiced signals are approximately repetitive on the pitch period; formants modify the shape of the speech wave, and require a frequency domain representation. Friction is aperiodic with spectral energy concentrated at higher frequencies. Variation of features is observable from consecutive frame-to-frame differences, and the range of values over a cluster of frames; more rapid variation ('transience') requires a more sophisticated mathematics to model non-stationarity of the signal. An important property of speech is the independent variation of F0, F1, F2 and F3.

Problems of accuracy arise in mathematical models of these observables in the presence of noise. Loss of high frequencies makes friction readily confusable with non-speech noise bursts. Periodic noise is confusable with voicing, especially if there is frequency variation. Robust mathematical models of non-stationary spectra are problematic. On the whole, speech is identifiable owing to variation of parameters, rather than constant effects; there are no simple invariants.

3.4   COMPUTATION ALGORITHMS

For the above features, the following are proposed as the most cost-effective techniques of estimation.

3.4.1   Energy estimation
It is desirable to compute energy in the frequency band expected for speech. Contrasting this speech-energy estimate with overall energy gives an estimate of noise content.

Measure AM (average magnitude) for overall energy is defined as

$$AM = \sum_{r=0}^{N-1} |x_{n+r}| \qquad \text{(normalised for } N\text{)}$$

where N is the frame length, typically 256 samples.

Cost saving is possible by downsampling (alternate samples) and also by omitting alternate frames. Summation can be kept in 16-bit range by using a shift operation to reduce the range of the samples.

Measure AM1 for speech energy is the same calculation applied to a smoothed signal, e.g.

$$AM1 = \sum_{r=0}^{4r<N} |y_{n+4r}| \qquad \text{(normalised for } N\text{)}$$

where $y_n = x_n + x_{n-1} + x_{n-2} + x_{n-3}$. Downsampling (4:1) is used here to reduce cost.

3.4.2 Periodicity estimation

The concepts of periodicity [4] and period cannot be separated. Assessing periodicity involves a search within the space of expected pitch periods. Some measure of the 'salience' of each period is applied. In speech, the pitch period is likely to be in the range 50–800 Hz. Pitch extraction is not straightforward, but is easier if one is prepared to tolerate errors of doubling.

Two low-cost techniques are available, somewhat complementary in effect. Both use comb filters tuned to the period being tested. Low-pass pre-filtering of the signal improves detection.

AMDF (average magnitude difference function) for period k,

$$AMDF\ (k) = \sum_{r=0}^{N-1} |n_{n+r} - x_{n+r+k}|$$

is an anticorrelation measure with lag k.

TDPA (time domain periodogram algorithm) first applies an enhancing filter, such as

$$y_n(k) = x_n + x_{n+k} \quad \text{or} \quad y_n(k) = x_n + x_{n+k} + x_{n+2k}$$

and tests the range of y over the frame, effectively responding to repeated prominences:

$$\text{period TDP} = \operatorname*{argmax}_{k} \left(\operatorname*{range}_{r=0}^{r<N} y_{n+r}\ (k) \right)$$

as an approximate maximum-likelihood method.

Neither technique is reliable during energy variation. AMDF is improved by centre- and peak-clipping, e.g. by severely quantising the signal.

The most useful measure of periodicity (regularity) is probably the AMDF minimum, normalised by energy. The TDPA itself gives a measure of energy rather than periodicity.

3.4.3   Spectral shape estimation

The discrete Fourier transform provides a short-term spectrum at a cost considerably greater than the above calculations. Efficient FFT implementations can reduce the cost to below eight multiplications per sample, or less with Winograd FFT [5].

Since the efficiency is achieved by simultaneous calculation of power in all frequencies (harmonics of the frame frequency), no saving is possible by directly tracking spectral peaks.

Linear prediction methods [6] are even more expensive, 30 to 100 multiplications per sample being typical. Kalman Filters [7] are a further predictive technique.

Simple time domain estimators of first and second formants, using Zero Crossing Intervals (ZXI), are feasible only in low noise. The smoothed signal input to periodicity algorithms should be dominated by the first formant, so that the short-term maximum ZXI will sometimes provide an estimator.

Non-stationary spectra

Estimation of transient spectra at consonant–vowel (C–V) boundaries is difficult. Three methods are encountered:

Wigner–Ville spectra [3] are designed to model the ear's response to transient events. Calculation is similar in cost to FFT.

Time-varying LPC models have been investigated in [8]; backward time-variant LPC is another approach [9].

Strategy and sequence of application

(1) AM analysis of the whole block of samples to be analysed.
(2) Test successive frames against an energy threshold. For those above threshold, use AM1. Segment the block into potential speech events ('syllables') according to predominance of AM1.
(3) Select frames with static AM1 for periodicity analysis, from within the syllables; at least one from each.
(4) Use an AMDF estimator on each frame after smoothing. Store pitch period estimates. Reject aperiodic frames.
(5) Track period forwards on continuity assumptions, using TDPA for period estimator, followed by AMDF to check periodicity. Revise syllable boundaries where AMDF fails. Store pitch range within syllables.
(6) For syllables with good periodicity, track pitch backwards to discover consonant boundaries.
(7) Track F1 in good syllables; track F2.
(8) Carry out non-stationary analysis on possible C–V boundaries.

Speech identification and quality estimation
The likelihood that a portion of the signal contains speech can be estimated by counting the number of speech-like events in the portion which have succeeded at a given level of the hierarchical analysis. An overall score can be calculated using suitable weights; consideration of weightings will depend on future experiments with signals of graded quality.

Tracking methods
Pitch tracking is carried out assuming continuity. Let $k(m)$ be the period estimate for frame m. Then the search interval for the next frame is $[k(m) - a, k(m) + a]$ for some suitable small value a, dependent on acceptable pitch variation.

Implementation
Algorithms have been implemented in 'C' for reasons of portability and efficiency of compiled code.

3.5   DISCUSSION

3.5.1   Model refinement
The hierarchical organisation can be viewed as a sequence of progressively refined models of speech, ranging from presence of energy to coherent rapid spectral variation at boundaries. Success at one level of the model is a criterion for testing to proceed at the next.

3.5.2   Limitations
Owing to contextual information, speech may be intelligible to a human listener where there is very little acoustic evidence. The overall model cannot take account of this; one could devise accurate algorithms for analysing such signals, deeply buried in noise, but this is quite infeasible where *a priori* information on presence of speech is absent. The intelligibility of such a buried event is liable to be most influenced by the nature of neighbouring events. Therefore such robust algorithms are far less use in the current problem than in, for instance, a recognition application.

Perhaps the most important problem with estimation of intelligibility is the need for a simple way to detect transient formants at C–V boundaries, which appears to be the best way to ensure that consonants are audible.

The algorithms presented include approximations which are expected to lead to their failure in specific conditions, but they should only fail systematically in very adverse noise conditions. Extensive experimentation will be required to discover how well the model proposed can be made to agree with the human listener's assessment. The structured approach enables new algorithms to be added or substituted to refine the model.

3.6   CONCLUSION

This chapter has proposed a hierarchical approach to low-cost speech quality estimation. Useful algorithms have been proposed where they exist, and areas of difficulty described. The model proposed is expected to provide a good basis for experimentation and framework for further improvements.

ACKNOWLEDGEMENTS

We wish to thank Dr Sohrab Saadat for his moral support for this research; also GlosCAT students Richard Bates, Russell Dutton and Paul Hornby for their work in implementation of several FFT and LPC algorithms in 'C'. We gratefully acknowledge the advice and support of John Haseler and colleagues at GCHQ Cheltenham.

REFERENCES

[1] G. R. Dattatreya & V. V. S. Sarma, 'Decision tree design and applications in speech processing', *IEE Proc.*, **131**, part F, No. 2, pp 146–152, Apr. (1984).

[2] J. L. Flanagan, *Speech Analysis and Perception*, (1983) Springer-Verlag, New York.

[3] David Chester, Fred J. Taylor & Mike Doyle, 'The Wigner distribution in speech processing applications', *J. Franklin Inst.*, **318**, No. 6, 415–430, Dec. (1984).

[4] Wolfgang Hess, *Pitch Determination of Speech Signals*, (1983) Springer-Verlag, New York.

[5] S. Winograd, 'On computing the discrete Fourier transform', *Maths Comp.*, **32**, No. 141, 175–199, Jan. (1978).

[6] J. D. Markel & A. H. Gray Jr., *Linear Prediction of Speech*, (1976) Springer-Verlag, New York.

[7] G. Carayannis, D. Manolakis & N. Kalouptsidis, 'Fast Kalman type algorithms for sequential signal processing', *IEEE Int. Conf. ASSP Proc.*, Boston, 186–189 (1983).

[8] Yves Grenier, 'Time-dependent ARMA modeling of nonstationary signals, *IEEE Trans.*, **ASSP-31**, No. 4, 899–911, Aug. (1983).

[9] Takayuki Nakajima & Torazo Suzuki, 'Non-steady state speech analysis by time-variant backward linear prediction', *Bulletin of Electrotechnical Laboratory Japan*, 6–11 (1983).

4

Further links between signal processing and channel coding with special reference to optical communications

John J. O'Reilly and **Manuel Monteiro**
School of Electronic Engineering and Science, University College of North Wales,
Dean Street, Bangor, Gwynedd, LL57 1UT

4.1   INTRODUCTION

Recent research has established close links between the two previously disparate disciplines of signal processing and error control coding [1,2]. The key unifying concept has been the use of Fourier transformations from a Galois field to an appropriate extension field with a suitably chosen primitive element as kernel of the transformation. Finite field transforms can be shown to provide a basis for error control encoding, with the inverse transform providing the starting point for the error detection, location and correction operations involved in decoding, in which there is an analogy with spectral estimation.

The same general ideas apply to any field, including R and C, and the familiar discrete Fourier transform (DFT) can then provide a means of achieving error control for real number sequences [2–4]. To date, though, such real number codes as have been reported have been cast in a largely theoretical setting devoid of potential application to any specific physical channel.

In this chapter the use of signal processing techniques to provide real number error control for analogue optical communications is considered. Specific attention is given to optical pulse position modulation [5], a twisted modulation scheme [6] in which samples are conveyed with generally high integrity save for occasional large distortions which may be identified as erasures. A class of codes is introduced, based on the DFT and also on the discrete Hartley transform (DHT) [7], which provides an efficient means of achieving error control to combat these erasures whilst providing positional disparity control to facilitate synchronisation.

4.2   CONVENTIONAL VIEW OF ERROR CONTROL CODING

A binary message vector of k bits may be augmented by $(n - k)$ parity bits to form an n-bit codeword. Errors which occur in transmission may then be detected and corrected provided the number of errors in a word is less than one half of the minimum Hamming distance between valid codewords. A simple example is provided by Fig. 4.1, which represents a 3-bit repetition code.

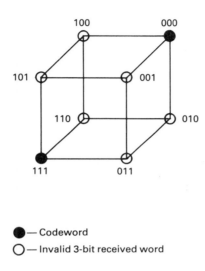

● — Codeword
○ — Invalid 3-bit received word

Fig. 4.1 — Simple example of 3-bit repetition code.

Let 0_0_0 and 1_1_1 be codewords. If, for example, 0_0_1 is received, we decode this as 0_0_0 assuming that only one error occurred rather than the two errors which would have been required to translate 1_1_1 into 0_0_1. We are making use here of a simple minimum distance decoding strategy.

4.3   THE HAMMING (7,4) CODE

A further example is the familiar Hamming (7,4) code. Here, four message bits are augmented by three parity bits to produce 7-bit codewords. These are 2^4 codewords. Any of the 2^7 possible received 7-bit words can be decoded into one of 16 codewords on the assumption that no more than one error has occurred. This is a single 'error detection and correction' code.

The 2^4 codewords form a subspace in the vector space of 2^7 possible 7-bit words. The underlying algebraic structure of this linear code facilitates both codeword generation and decoding.

The usual mathematical representation is in terms of a polynomial over a finite field, in this case GF(2).

4.4   LINKS BETWEEN ERROR CONTROL CODING AND SIGNAL PROCESSING

Recently, error control coding has been interpreted in terms of Finite Field Fourier Transforms or Galois Field Transforms (GFTs) [1]. This provides a link with signal processing. The essential idea may be appreciated by reference to Fig. 4.2, as

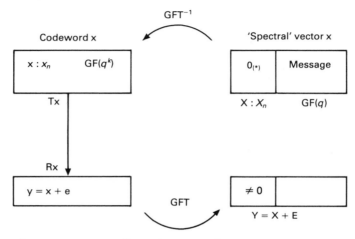

e is a vector corresponding to the pattern of errors

(*) are the parity 'frequencies'

Fig. 4.2 — Error control coding and signal processing.

follows. Let **X** be a spectral vector, containing the data augmented by zeros which represents the redundancy, referred to as parity frequencies. The inverse Galois field transform of **X** yields **x**, the corresponding (non-systematic) codeword. With **X** defined over $GF(q)$, **x** will in general be defined over some extension field, $GF(q^k)$. The codeword **x** is transmitted and may be corrupted so that the received vector is **y** = **x** + **e**, where **e** represents the pattern of errors. The GFT of **y** yields a spectral vector **Y** = **X** + **E**, where **E** is the error spectrum. Part of **E** is known — those elements of **X** which were set at zero and are now not zero — the 'parity' frequencies.

Spectral estimation [1] can now be used to determine the rest of **E**, the constraints originating from the limited number of errors given by the non-zero component in **e**.

Hence, **X** = **Y** − **E** enables the data to be recovered.

We note that:

(1) While the data are embedded in a vector defined over $GF(q)$, the resultant codeword may be defined over an extension field $GF(q^k)$.

(2) The form of the Finite Field Fourier Transform is essentially the same as that of the familiar Discrete Fourier Transform except that instead of using as a kernel $\exp(-i2\pi/N)$, the Nth root of unity in the field of complex numbers, we employ

α, a primitive element of GF(q^k) such that $α^N = 1$ but $α^r ≠ 1$ for any r such that $0 < r < N$.

(3) The ideas apply to any field, not just finite fields. They are thus applicable in principle to R and C, and may provide a means of achieving error control for pulse transmission systems.

4.5 PULSE POSITION MODULATION

Analogue or digital signals may be conveyed by pulse position modulation (PPM). The transmitted signal comprises of data 'frames', each containing a single pulse with the position of the pulse in a frame representing the value of the corresponding message sample.

For analogue transmission there is a continuum of possible pulse positions per frame whereas for digital PPM pulses these can occur at only certain discrete time instants. An illustrative segment of a digital PPM signal is shown in Fig. 4.3.

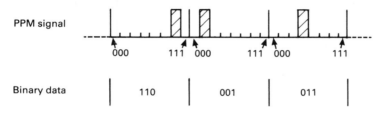

Fig. 4.3 — Illustration of digital PPM.

4.6 THE OPTICAL PPM CHANNEL

We note that PPM has been shown to be a particularly appropriate signal format for certain optical communications applications [5,8]. It has the potential merit that the average optical energy required per bit of information conveyed can be very low — in principle, fractions of a photon per bit!

Under these circumstances, though, there is a significant probability of signal pulses not being detected, so that a data frame may be received with no pulse in it, as illustrated in Fig. 4.4. We refer to this as an erasure.

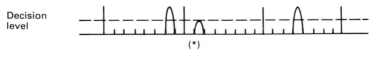

(*) — Pulse not detected → erasure

Fig. 4.4 — Optical PPM channel.

Note that the occurrence of an erasure is immediately apparent: there is thus merit in identifying for the optical PPM channel coding schemes which can provide for erasure correction.

4.7   SYNCHRONISATION AND LINE CODING FOR PPM

In order to interpret correctly a received PPM signal, we must be able to produce a reference signal synchronised to frame boundaries. This can be achieved by means of a Phase-Locked Loop (PLL), which adapts to the short-term pulse repetition frequency and phase of the PPM signal. It is, of course, essential that the PLL should track any long-term drifts in phase or frequency, but it should not track short-term phase variations corresponding to the modulation process. It follows that we must ensure that the message signal does not contain d.c. and low-frequency components.

For digital PPM, suppression of d.c. and low-frequency components can be achieved by making use of a digital line code. Redundancy is introduced into the message so that the short-term average location of pulses in the PPM signal is maintained near the centre of the frame. A sequence of 'early' pulses, for example, would be compensated for by subsequent 'late' pulses. Digital PPM line codes which facilitate synchronisation with very high coding efficiency ($\approx 90\%$) have been reported recently in the literature [9] and their synchronisation capabilities assessed [10]. For analogue PPM it may be the case that the message does not contain significant low-frequency and d.c. terms making direct use of a PLL for synchronisation possible. Where this is not the case, difficulties can arise. As we shall now see, though, it is possible to introduce line coding for analogue sampled signals to facilitate synchronisation and, moreover, to simultaneously achieve erasure control coding.

4.8   COMBINED LINE CODING AND ERASURE CONTROL FOR
ANALOGUE PPM TRANSMISSION

We recall our previous discussion of error control coding viewed as essentially involving Fourier Transforms. Conventionally, the vector components are considered to be elements of a finite field but, as was pointed out, these ideas apply to any field, including R and C. We noted that redundancy could be introduced by appending zeros to a data vector and viewing this as a 'spectrum', the transformation of which yields the desired codeword. It follows that if we make use of DFT, working with elements in R and C, and introduce redundancy by assigning the value zero to the zero frequency component of the spectral vectors, the analogue codewords obtained by inverse Fourier transformation will have zero d.c. content. In transmitting a sequence of such codewords over a PPM channel, we can be assured that successive pulse locations will, on average, group themselves at the centre of the frame. It follows that a simple PLL synchroniser can then be employed.

That this scheme can also provide for erasure control can be deduced from our previous discussion of transform techniques for error control, and is readily illustrated, as shown in Fig. 4.5.

4.9   CODING FOR R AND C

We may base analogue erasure control coding on, for example:

(1) The Discrete Fourier Transform with:

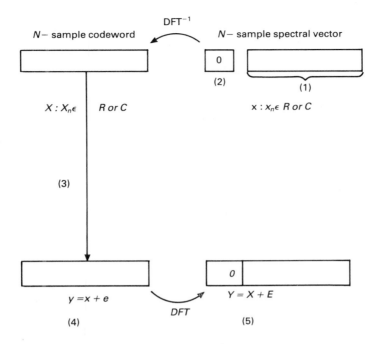

Fig. 4.5 — Line coding and erasure control. (1) $(N-1)$ analogue data samples, (2) zero frequency term $= 0$, (3) erasure (known location) may occur in transmission), (4) e defines the location of an erasure, (5) the value of the zero frequency term gives the 'correction' required for the erasure enabling X to be determined.

$$X_n \in C$$
$$x_n \in C$$

These complex samples can therefore be transmitted as pairs of PPM frames.
(2) The Discrete Fourier Transform embedding the data in **X** to ensure Hermitian Symmetry, so that **x** is defined over R, obviating the need for pairing PPM frames.
(3) The Discrete Hartley Transform

$$\mathbf{X} : X_n \in R$$
$$\mathbf{x} : x_n \in R$$

The zero frequency term of the DHT corresponds to the zero frequency term of the corresponding DFT, so that setting the former to zero ensures zero d.c. for the PPM signal.
These discrete transforms are given by:

The DFT

$$F(m) = \frac{1}{N} \sum_{n=0}^{N-1} f(n)\, e^{-i2\pi mn/N}$$

$$f(n) = -\sum_{m=0}^{N-1} F(m)\, e^{+i2\pi mn/N}$$

The DHT

$$H(m) = \frac{1}{N}\sum_{m=0}^{N-1} h(n)\, \mathrm{cas}(2\pi mn/N)$$

$$h(n) = -\sum_{m=0}^{N-1} H(m)\, \mathrm{cas}(2\pi mn/N)$$

where $\mathrm{cas}\,\theta = \cos\theta + \sin\theta$.

In each case, fast algorithms exist for suitably chosen values of N [11,12].

4.10   DFT AND DHT CODING ILLUSTRATIONS

The application of DFT and DHT coding to analogue PPM transmission is illustrated in the examples presented in Tables 4.1 and 4.2. In each case, a zero d.c. term is introduced to effect line coding to facilitate synchronisation, while providing for erasure correction as a form of error control especially appropriate for the optical PPM channel.

4.11   CONCLUDING REMARKS

We have shown that signal processing techniques can provide a useful basis for real number error control appropriate to analogue PPM transmission. Simple erasure control codes have been illustrated based both on the familiar discrete Fourier transform and on the discrete Hartley transform. The latter has the operational merit of mapping real number sequences into real number sequences, obviating the need for additional manipulations associated with use of the DFT to map complex vectors onto the real number sequence of the PPM channel. We note that although we have concentrated here on erasure control, it is possible to achieve both erasure and error correction by introducing further redundancy.

The results presented provide an illustration that, by building upon signal processing concepts, channel codes may be designed to achieve simultaneously the previously separate functions of error control and line coding [13]: a topic deserving further investigation in which the synergy of signal processing and error control techniques can be expected to prove to be particularly beneficial.

ACKNOWLEDGEMENTS

This research was supported by a grant from the UK Science and Engineering Research Council.

Table 4.1 — DFT line coding, without Hermitian symmetry, example.

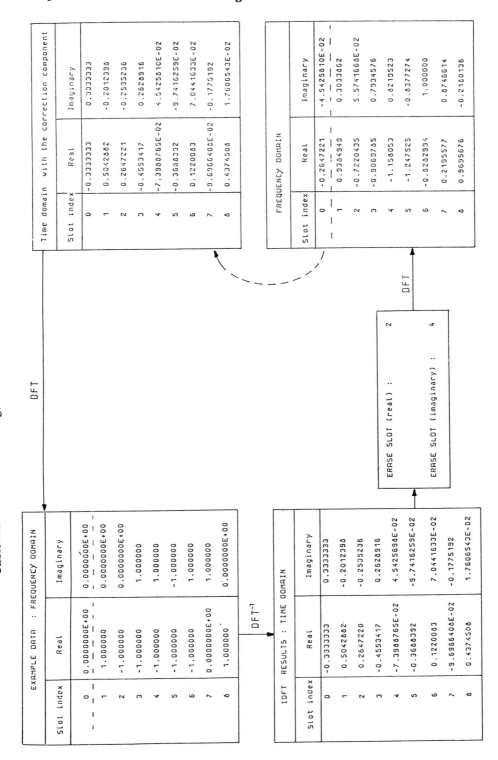

Table 4.2 — DHT line coding example.

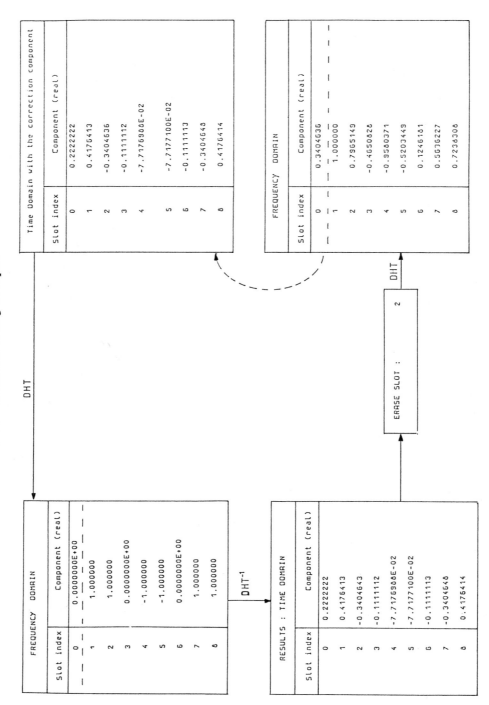

REFERENCES

[1] R. E. Blahut, 'Transform techniques for error control codes', *I.B.M. J. Res. Develop.*, **23**, No. 3, 229–315, May (1979).

[2] J. J. O'Reilly, 'Finite field Fourier transforms and error control coding', *Digest of Colloquium*: *Mathematical Topics in Telecommunications — Algebraic Structures and Error Control Coding*, University of Essex, pp. 6.1–6.13, January (1984).

[3] J. K. Wolf, 'Redundancy and the discrete Fourier transform', *Proc. Conf. Inf. Sci. Syst.*, 156–158, March (1982) Princeton, USA.

[4] T. G. Marshall, 'Coding of real-number sequences for error correction: a digital signal processing problem', *IEEE JSAC*, **SAC-2**, No. 2, 381–392, March (1984).

[5] I. Garratt, 'Pulse position modulation for transmission over optical fibres with direct or heterodyne detection', *IEEE Trans.*, **COM-31**, No. 4, 518–527, April (1983).

[6] J., M. Wozencroft & I. M. Jacobs, *Principles of Communication Engineering*, p. 613, (1965) Wiley, New York.

[7] R. N. Bracewell, *The Hartley Transform*, Chapter 4, p. 27, (1986) O.U.P.

[8] S. D. Personick, 'Application of quantum estimation theory to analogue communication over quantum channels'. *IEEE Trans. Inf. Theory*, **IT-17**, 240–246, May (1971).

[9] J. J. O'Reilly & Y. Wang, 'Line code design for digital pulse position modulation', *Proc. IEE, Pt.F*, **132**, No. 6, 441–446, Oct. (1985).

[10] Y. Wang & J. J. O'Reilly, 'Synchronization of line-coded digital PPM in repeated transmission systems', *Proc. IEE, Pt. F* (in press).

[11] R. E. Balhut, *Fast Algorithms for Digital Signal Processing*, (1985) Addison-Wesley, Reading, MA.

[12] See [7] Chapter 8, p. 79.

[13] J. J. O'Reilly, 'Line coding for disparity and spectral control', *Digest IMA Conf. Cryptography and Coding*, December (1986) Cirencester, UK.

5

Minimisation of architectural complexity in the design of adaptive equalisers for digital radio systems

K. V. Lever
School of Information Systems, University of East Anglia
F. M. Clayton
GEC Research Ltd, Telecommunications Laboratory, Hirst Research Centre, Wembley, Middlesex, HA9 7PP, UK
G. J. Janacek
School of Mathematics and Physics, University of East Anglia, University Plain, Norwich, NR4 7TJ, UK

5.1 INTRODUCTION

Modern high-speed digital communication systems often make use of terrestrial radio links for part or all of their interconnection. Such links carry multi-level digital signals, with sample-rates typically in the region of 20 to 75 million symbols per second, modulated on to sinusoidal carrier frequencies typically in the region of 2 to 20 GHz.

In normal conditions, the signal arriving at the receiving aerial dish is undistorted by its passage through the atmosphere. Occasionally, however, particular atmospheric conditions will permit more than one propagation path between transmitter and receiver. One such mechanism is anomalous refractive-index variation with height often associated with stable anticyclones as depicted in Fig. 5.1. This gives rise to the propagation of an anomalous ray in addition to the normal ray as shown. If the propagation paths differ in length by an odd number of half-wavelengths (and at the frequencies of interest, this equates to only a centimetre or two), then the mechanism of destructive interference will come into play.

The resultant distortion of the communication channel is linear, but frequency selective: attenuation and group-delay are different at different frequencies, causing redistribution of energy in the digital signal (which necessarily occupies a wide region

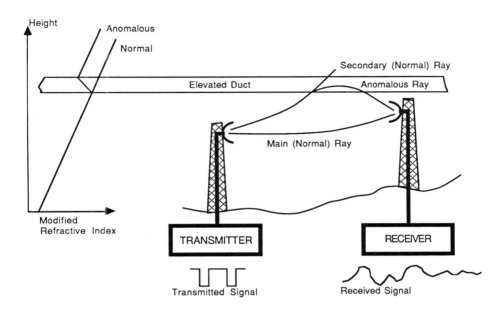

Fig. 5.1 — Anomalous refractive-index variation and multipath propagation.

in the frequency domain). Consequently, by the time the signal is demodulated down to baseband (that is, stripped of its carrier-frequency content), what started out from the transmitter as a series of clearly-defined digital pulses confined to their own signalling interval arrives at the receiver in a very sorry state indeed: attenuated, distorted and dispersed over several neighbouring signalling intervals. The signal is said to suffer **intersymbol interference** (ISI).

Any equaliser designed to remove these distortions must effectively act in a complicated frequency-dependent fashion that restores the original channel characteristics. Note that since these systems employ modulation of both the amplitude and the phase of the carrier (that is, the transmitter sends **complex** rather than real numbers over the channel to the receiver), the equaliser has to correct both the real and the imaginary parts of the received signal. Such systems are necessarily rather complicated two-channel signal processors: one channel to handle the real (inphase) signal component, the other to handle the complex (quadrature) component. In addition to all this, the channel is **time-varying** (for example, atmospheric refractive-index stratifications may slowly change their vertical position) so that the equaliser is required to **adapt** to the channel variation. It must do this without the benefit of a 'side channel' and without interrupting the transmission of data: it must diagnose the channel characteristics from observations it makes on the distorted signals it receives. Clearly, the equaliser must perform the estimation/correction computations at a rate equal to the rate of variation of the channel. Helpfully, this rate is about four or five orders of magnitude lower than the symbol rate, so that the channel variation can be regarded as a secular perturbation: typically this can be dealt with fairly easily by means of control loops with appropriate bandwidth. The chief problem is designing equaliser architectures capable of coping with the very high symbol rates, that can be implemented with sufficiently high accuracy at an acceptably low cost. As always in commercial enterprise, cost-effectiveness is the watchword, and the designer naturally gravitates towards techniques that can be used to reduce the complexity of the equipment (without loss of performance): indeed the quest is for **minimal** complexity.

In what follows, the conventional representations of Fig. 5.2 are used, with signal flow in the direction of the arrowheads representing variable gain amplifiers (analogue) or multipliers (digital).

ADAPTIVE TAP   UNIT DELAY   SUMMATION

Fig. 5.2 — Signal processing modules.

Conventional approaches to equalisation exploit the fact that, though time-varying, the channel is **linear,** so that its effect on the transmitted signal can be modelled as a **convolution operator.** Accordingly, the equaliser is an embodiment, at least approximately, of thw **inverse deconvolution operator.** In view of the linearity of the distortion mechanism, it is enough to equalise the channel impulse response: the Principle of Superposition ensures that this implies equalisation for arbitrary input signal waveforms. Nearly all current designs, as summarised for example in [1], are based on a two-stage equaliser architecture in which **precursors,** signal samples preceding the main pulse, are dealt with by means of a **feedforward transversal** processor (tapped delay line), while the **postcursors** are removed by means of a **feedback** processor, as shown in Fig. 5.3.

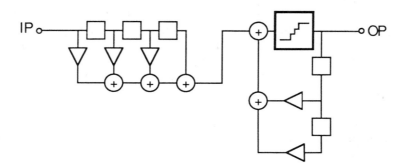

Fig. 5.3 — Decision feedback equalisation.

Note the presence of **decision** feedback, in which the regenerated multi-level digital signal, rather than an analogue signal, is fed back. Though this renders the operation non-linear, it can be shown that this is beneficial in removing some of the unwanted noise accompanying the wanted signal, and in ensuring the stability of the feedback loop.

We shall no longer concern ourselves with the feedback subsystem, but will consider further the case depicted in Fig. 5.2, in which the feedforward subsystem contains three adaptively variable taps plus one (final) unity-gain tap. Such a processor has three degrees of freedom which, under closed-loop control, may be disposed to counteract precursor intersymbol interference. We consider one of the simplest approaches, in which the three precursors are forced to zero by the control algorithm (see Fig. 5.4).

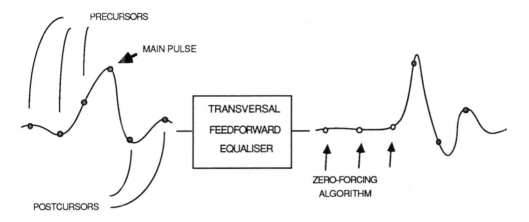

Fig. 5.4 — Precursor equalisation.

The question of whether this three-lap fully-transversal equaliser can be simplified has recently been studied by a team of engineers at the Telecommunications Laboratories of GEC Research Ltd, Hirst Research Centre, under the leadership of M. T. Dudek [2]. Assuming that the intersymbol interference is dominated by a **single** precursor, they have designed an economised version of the three-tap architecture employing only two taps, in what they call the **Series Architecture** (as shown in Fig. 5.5).

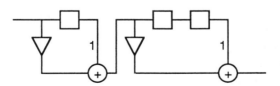

Fig. 5.5 — Two-tap series equaliser, first configuration.

When the performance of this series equaliser is measured experimentally in laboratory conditions simulating two-path fading [3], it is found to be very close to, perhaps a little better than, that of its three-tap **Standard** counterpart (as the conventional fully-transversal form is called).

There is clearly a second equivalent configuration (shown in Fig. 5.6) which can be obtained merely by reversing the order of the two subsystems: linear convolutional operators necessarily commute.

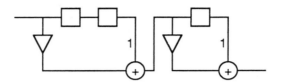

Fig. 5.6 — Two-tap series equaliser, second configuration.

Though trivially related to the first, this second configuration is considered to be a **distinct** architecture from the point of view of patent protection: we shall return later to the question of the **number of configurations** (two in this case) as distinct from the **number of designs** (one in this case).

Notice that there is no assumption that the single-precursor model always holds good. On the contrary, multi-precursor ISI is often observed. All that is being claimed is that the two systems described above have a measured performance for two-path fading which is very close to that of the three-tap standard equaliser (illustrated in Fig. 5.7).

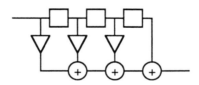

Fig. 5.7 — Three-tap standard equaliser.

The question then naturally arises: can the same decomposition be applied to, say, a two-tap or four-tap standard equaliser? What determines whether such a decomposition is possible? How does the designer know when no further economisation is possible? In the next section we present a theoretical model for the decomposition process, and develop a design technique based on combinatorial optimisation that leads to a clearer understanding how a Standard equaliser may be reduced to its simplest Series equivalent.

5.2   THE SINGLE-PRECURSOR SIGNAL AND THE SERIES EQUALISER

We start by considering the signal model shown in Fig. 5.8 in which the main pulse is, without loss of generality, taken to have unity magnitude; and the precursor is taken to have magnitude $-b$ (the minus sign prevents the notationally awkward proliferation of alternating $+$ and $-$ in the equaliser impulse response).

The signal to be equalised can be represented as:

$$x = [-b, 1]$$

The impulse response of the standard three-tap equaliser designed to force three zeros is

$$h = [b^3, b^2, b, 1]$$

The output signal from the equaliser is the discrete convolution

$$y = h*x = [-b^4, 0, 0, 0, 1]$$

so that the residual intersymbol interfeence (RISI) is of fourth order:

$$\text{RISI}(4) = -b^4$$

The two-tap series equaliser can be shown to provide **precisely** the same performance. The two tap coefficients will adapt to the values b and b^2 respectively to force two zeros, but a third 'free' zero is also obtained. The reason is that the convolution of the impulse responses of the two subsystems gives exactly the impulse response required to force three zeros:

$$[b, 1]*[b^2, 0, 1] = [b^3, b^2, b, 1]$$

It is as if the series architecture synthesises a third 'virtual' tap of exactly the required magnitude. Note that the series configuration is a third-order system (it has three delays) despite the fact that there are only two adaptive taps. One degree of freedom has been omitted, and the second subsystem shows the 'thinning' (omission of the middle tap) to which the invention owes its economy. Thus, for a single-precursor ISI model, the Series Architecture appears to be capable of resynthesising the missing tap. We now consider the general case of an N-tap equaliser.

5.3   FORMAL DESIGN USING A LATTICE MODEL

Our design objective is, for the general case, to design an equaliser, H, which maps the single-precursor input signal $x = [-b, 1]$ into another single-precursor signal, the output, in which the residual ISI is of order n, by forcing $N = n - 1$ zeros:

$$H = [-b, 1] \rightarrow [-b^n, \underbrace{0, 0, ..., 0}_{N \text{ zeros}}, 1]$$

Notice that the action of the equaliser is to make the interference smaller. Indeed, n may be large enough to ensure that RISI(n) is negligible.

The structure of the required equaliser operator is convolutional, and one solution to the problem is the conventional standard N-tap equaliser:

$$[-b, 1]*[B_N, B_{N-1}, ... \, B_2, B_1, 1] = [-b^n, 0, 0, ..., 0, 0, 1]$$

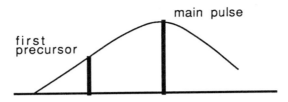

Fig. 5.8 — Single-precursor signal model.

This architecture, shown in Fig. 5.9, has the N degrees of freedom required to force N zeros: but some of these may be unnecessary, as the following analysis shows.

We use the z-transform [4] to convert the convolution to multiplication. Since z^{-1} represents a delay of one sample, z represents the 'anticipative' (negative) delay with respect to the main pulse appropriate to precursor structure. We have:

$$(-bz + 1) (B_N z^N + B_{N-1} z^{N-1} + \ldots + B_2 z^2 + B_1 z + 1) = (-b^n z^n + 1)$$

so that

$$B_1 = b, \quad B_2 = b^2, \quad \ldots, \quad B_{n-2} = b^{n-2}, \quad B_{n-1} = b^{n-1}$$

The standard equaliser structure having tap values forming a truncated geometric progression corresponds to a direct representation of this solution: an obvious generalisation of the three-tap case examined in the previous section. Other representations of this solution are possible, however, and it is these that correspond to the alternative Series architecture. We simplify the notation by writing $x = bz$.

Since $1 + x + x^2 + \ldots + x^{n-1}$ represents the transfer function of a linear system (the equaliser), we follow customary practice and seek its factors. These, together with $x = 1$ corresponding to the input signal model $1 - x$, are given by the roots of the equation

$$x^n = 1$$

that is, the n distinct nth roots of unity:

$$X = \left\{ x_k \right\}_{k=0}^{n-1}$$

where

$$x_k = \cos(2\pi k/n) + j\sin(2\pi k/n)$$

So we have

$$\sum_{k=0}^{n-1} x^k = \prod_{k=1}^{n-1} (x - x_k)$$

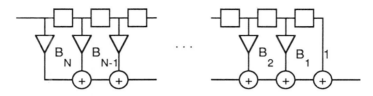

Fig. 5.9 — Standard N-tap equaliser.

Our problem would appear to be solved: the Nth-order equaliser can be decomposed into a cascade of $n - 1 = N$ first-order sections as in Fig. 5.10.

In a sense, this **is** a Series architecture in that it exhibits a modular structure comprising a series of subsystems connected in tandem. At first sight, this decomposition would appear to require precisely as many taps as the fully-transversal standard equaliser of Fig. 5.11.

Closer scrutiny reveals, however, that in the general case when all (non-unity) roots are complex, the taps would each have to be **complex,** equivalent to two real taps. Thus the architecture of Fig. 5.11 requires double the number of taps of the standard architecture. This is unsatisfactory, as our task is to find an architecture of **reduced** complexity: indeed our aim is to **minimise** complexity.

The reason for the complexity of the first-order cascade is patent: we are allowing the roots to lie in the complex field ($x_k \in \mathbb{C}$). There are accordingly always N roots, giving rise to N equaliser zeros and N equaliser sections. Notice that there is no *arithmetic* problem in allowing complex roots. Rather, the problem is *architectural*: the design requires the full complement of N complex taps, and we would prefer to use fewer real ones. It is possible that we might reduce the number of taps by reducing the number of factors of $x^n - 1$. This can be done by combining the monomial factors into binomials or trinomials or higher-order polynomials, by taking pairs or triples or more factors together and expanding the brackets. This will only help if it leads to some cancellation of middle terms in the polynomials that arise: zero coefficients correspond to taps that can be omitted. In fact it is possible to group factors together in order to achieve a reduction. What is more, there are a number of ways in which one can set about doing this in a systematic fashion.

One method makes use of the fact that the n nth roots of unit form a **Cyclic Group of order** n, $\mathbb{C}_n$, and that its elements can be sorted into equivalence classes, for grouping as above, in a natural way using the concept of **Cyclotomy** [5,6]. The equaliser transfer function of order N can be written as a product of appropriate Cyclotomic polynomials. These can then be grouped so as to minimise the number of coeffients in the resultant product of higher-order polynomials. To the extent that such grouping is necessary, the approach is indirect. For the purpose of presenting the design technique in as straightforward a fashion as possible, we prefer an equivalent approach which makes use of Group Theory in a different way, giving the required designs in a completely direct fashion.

It turns out that we can obtain some simplification of the expression $1 + x + x^2 + \ldots + x^{n-1}$ (in the sense that fewer taps are needed for implementation) by examining the **subgroup structure** of $\mathbb{C}_n$. If n is prime (having no divisors save itself and 1) then

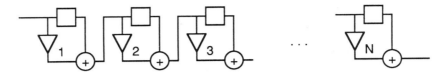

Fig. 5.10 — Cascade of first-order sections.

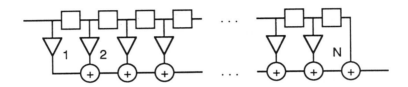

Fig. 5.11 — Standard transversal architecture.

$\mathbb{C}_n$ will have no subgroups (save the trivial cases $\mathbb{C}_n \subset \mathbb{C}_n$ and $\mathbb{C}_1 \subset \mathbb{C}_n$). But if n is composite then subgroups exist and it is a simple matter to enumerate them all, for all the cases of small n likely to be of practical interest. Indeed this task has already been done, and tables of group structure are readily available†, for example [8,9]. Because subgroup inclusion is a partial ordering, the set of subgroups forms a **lattice** [10] (with binary operations $\cup$ and $\cap$), and it is actually the structure of the subgroup lattice, $\mathcal{L}_n$, that gives rise to reduction in the architectural complexity of the equaliser.

In what follows we make use of the **symbolic composition operator,** $\square$, in a *formal* fashion. The intention is to adopt an *operational* approach to the design process, without paying too much attention to the exact definition of the operation *per se*. There is a certain sense in which it is intuitively obvious that, if n is composite (say $n = r.s$) and $\mathbb{C}_r$ and $\mathbb{C}_s$ are both subgroups of $\mathbb{C}_n$, then

$$\mathbb{C}_n = \mathbb{C}_r \square \mathbb{C}_s$$

Certainly, we can regard $\square$-composition as the direct (Cartesian) product, $\otimes$, and if we do we must accept that it is commutative. There is another sense, however, in which we choose to visualise $\square$-composition as non-commutative: $\mathbb{C}_r \square \mathbb{C}_s$ and $\mathbb{C}_s \square \mathbb{C}_r$ **induce entirely different decompositions** of $\mathbb{C}_n$. If we consider $\mathbb{C}_r$ and $\mathbb{C}_s$ as elements of a commutative algebra we are, in effect, working with the induced K-module [11]. Rather than make direct use of K-module theory, however, we will instead clarify the design procedure by means of an example.

We choose the case $n = 12$, because this is a practical configuration (indeed it

† The same information (and more besides) is also available in software form: for example, CAYLEY is an expert-system-like package for solving problems in Group Theory [7].

seems rather too large in that respect: the current limits of practical feasibility are of the order 8–10), but because 12 is highly composite for its size: $12 = 2^2.3$. The subgroup lattice is correspondingly rich in substructure, as shown in Fig. 5.12.

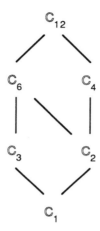

Fig. 5.12 — The subgroup lattice $\mathcal{L}_{12}$.

In Fig. 5.12 the descending lines denote proper inclusion. For notational unformity we have depicted the identity element, the only element common to all subgroups, at the foot of the diagram as C_1.

Clearly, C_{12} can be modelled in several ways. The first model is obviously C_{12} itself, whose elements (less the identity element x_0) form the roots of the 11th-order polynomial $1 + x + x^2 + \ldots + x^{11}$ corresponding to the Standard 11-tap transversal equaliser capable of forcing 11 zeros. Thus

$$C_{12} \leftrightarrow 1 + x + x^2 + x^3 + x^4 + x^5 + x^6 + x^7 + x^8 + x^9 + x^{10} + x^{11}$$

corresponding to the Standard architecture of Fig. 5.13.

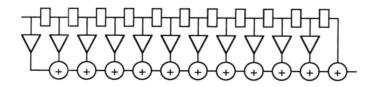

Fig. 5.13 — Standard 11-tap architecture.

But we also have the decomposition:

$$\mathbb{C}_{12} = \mathbb{C}_2 \,\square\, \mathbb{C}_6 \leftrightarrow (1+x)(1+x^2+x^4+x^6+x^8+x^{10})$$

corresponding to the six-tap Series architecture of Fig. 5.14.

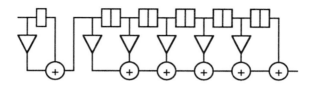

Fig. 5.14 — Series six-tap architecture, first form.

Notice that we can interpret these algebraic decompositions in a distinctly physical fashion: if we think of 6 as being 12 'thinned'† by a factor of 2, the architecture follows immediately. The first section is a one-tap module corresponding to $\mathbb{C}_2$, whereupon the following 5-tap module corresponding to $\mathbb{C}_6$ must be 'thinned' by the factor incorporated into the first section. Alternatively, we can think of the modular delay in the second module as double that of the first.

There is a second decomposition involving $\mathbb{C}_2$ and $\mathbb{C}_6$, in which their roles are reversed. This gives rise to an architecturally distinct design having the same complexity:

$$\mathbb{C}_{12} = \mathbb{C}_6 \,\square\, \mathbb{C}_2 \leftrightarrow (1+x+x^2+x^3+x^4+x^5)(1+x^6)$$

Note that the first factor corresponds to a Standard five-tap transversal equaliser, and the second to a one-tap module thinned by a factor of 6 as in Fig. 5.15.

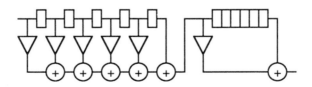

Fig. 5.15 — Series six-tap architecture, second form.

There are two distinct five-tap architectures, shown respectively in Figs. 5.16 and 5.17, derived from $\mathbb{C}_3$ and $\mathbb{C}_4$:

$$\mathbb{C}_{12} = \mathbb{C}_3 \,\square\, \mathbb{C}_4 \leftrightarrow (1+x+x^2)(1+x^3+x^6+x^9)$$

† The concept of 'thinning' is fairly common in Signal Processing, and was first encountered by the author during his work on Surface Acoustic Wave matched filter banks for Multi-frequency Shift Keyed communication systems [12,13].

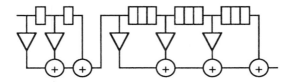

Fig. 5.16 — Series five-tap architecture, first form.

$$\mathbb{C}_{12} = \mathbb{C}_4 \,\square\, \mathbb{C}_3 \leftrightarrow (1 + x + x^2 + x^3)(1 + x^4 + x^8)$$

Finally, we consider the decompositions which correspond to the **standard form** (product-of-powers-of-primes) of the factorisation of 12, i.e. $12 = 2^2.3$. There are three distinct 4-tap architectures, corresponding to the three distinct ways two 2s and one 3 can be arranged. These are shown in Fig. 5.18, 5.19 and 5.20, respectively.

$$\mathbb{C}_{12} = \mathbb{C}_2 \,\square\, \mathbb{C}_2 \,\square\, \mathbb{C}_3 \leftrightarrow (1 + x)(1 + x^2)(1 + x^4 + x^8)$$

$$\mathbb{C}_{12} = \mathbb{C}_2 \,\square\, \mathbb{C}_3 \,\square\, \mathbb{C}_2 \leftrightarrow (1 + x)(1 + x^2 + x^4)(1 + x^6)$$

$$\mathbb{C}_{12} = \mathbb{C}_3 \,\square\, \mathbb{C}_2 \,\square\, \mathbb{C}_2 \leftrightarrow (1 + x + x^2)(1 + x^3)(1 + x^6)$$

Notice that thinning is cumulative: in these last three designs the cascade is composed of three modules, and the last module is thinned by a factor corresponding to the **product** of the factors of its predecessors. Thus we see that the more we factorise n, the more thinning can be achieved. **Since no finer resolution can be obtained than that corresponding to the standard form (in the product-of-powers-of-primes sense) then this form corresponds to the architectures of minimum complexity.** In particular, four taps are minimum for $n = 12$. We say that the Complexity is $C = 4$. There are, however, three distinct ways that this minimal architecture can be implemented — we say that there are three distinct designs: $D = 3$. In addition, since linear systems are commutative, the three modules in the cascade can be arbitrarily reordered without affecting the overall transfer function. There are $3! = 6$ ways of doing this, so that in all these are $3.6 = 18$ entirely equivalent configurations: we say $E = 18$.

The example given above illustrates the general principles required to determine minimal architectures for any case of interest. We have carried out this design process for all cases up to and including equaliser order $N = 31$ ($n = 32$). This is likely to cover all cases of foreseeable interest: and perhaps only the low-order cases ($n \le 8$) can be regarded as currently practicable. Accordingly, we detail in Table 5.1 the design of the low-order cases $N \le 7$ ($n \le 8$), deliberately adopting a connectivity representation similar to that found in [14] for feedback shift registers used in designing cyclic error correction codecs and m-sequence generators — except that we use a binary notation rather than octal. For brevity, only the **minimal** designs are recorded. Rather than using space-consuming diagrams to represent each architecture, we use binary sequences to indicate whether a tap is present (1) or absent (0). The separation of the modules in the design is indicated by a semicolon (;), and the

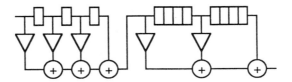

Fig. 5.17 — Series five-tap architecture, second form.

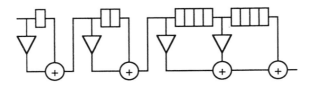

Fig. 5.18 — Series four-tap architecture, first form.

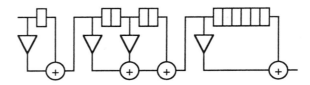

Fig. 5.19 — Series four-tap architecture, second form.

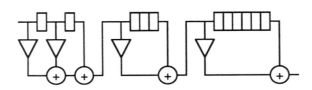

Fig. 5.20 — Series four-tap architecture, third form.

Table 5.1 — Low-order series equaliser designs

ORDER N	SPAN n	ARCHITECTURAL CONNECTIVITY	NO. OF TAPS C
1	2	1(1)	1
2	3	11(1)	2
3	4	1(1);10(1)	2
4	5	1111(1)	4
5	6	1(1);1010(1)	3
		11(1);100(1)	3
6	7	111111(1)	6
7	8	1(1);10(1);100(1)	3

final unity-gain tap is denoted as (1). The 'span' of the Nth-order filter is defined as $n = N + 1$. Thus these first cases can be abbreviated as in Table 5.1. In addition, we summarise in Table 5.2 a survey of the architectural characteristics for $N \leq 31$ ($n \leq 32$), without giving the details of the circuit topology for each case.

5.4 COMPLEXITY ANALYSIS

In this section we study the complexity of the minimal equaliser designs using ideas akin to those of Complexity Theory [15]. From this point of view we can regard the Series architecture as the hardware embodiment of a signal processing **algorithm,** and concentrate on the analysis of the algorithm. Practical considerations indicate that the cost of delay is small compared with that of an adaptive tap, so that the number of taps, C, is an accurate estimate of the total cost of the equaliser. For this reason we term C the **complexity** of the equaliser, and have orientated our design effort towards its minimisation. We can derive an algebraic expression for C as a function of n, the **span** of the equaliser (or equivalently, as a function of N, its **order,** the total number of delays). In addition, we can show that the Series algorithm reduces the complexity from $O(n)$ to approximately $O(\log_2(n))$ compared with the Standard fully-transversal algorithm. A reduction of this type is considered the paradigm of Complexity Theory: once the $n \rightarrow \log_2(n)$ reduction is achieved, it is unlikely that any further improvement will be forthcoming.

Consider the product-of-powers-of-primes **standard form**† of the factorisation of n:

$$n = p_1^{m_1} \, p_2^{m_2} \, \cdots \, p_k^{m_k}$$

This decomposition, as we have seen, is **optimal** in that it leads to **minimal** equaliser complexity. Compared with other possible factorisations, that using prime powers is

---

† Note the unfortunate clash of terminology: the **standard form** of Number Theory does not correspond to the Standard equaliser form.

Table 5.2 — Series equaliser designs

RISI Location n	No. of Standard Taps N	No. of Series Taps C	No. of Designs D	No. of Configurations E
2	1	-	1	1
3	2	-	1	1
$4 = 2^2$	3	2	1	2
5	4	-	1	1
$6 = 2.3$	5	3	2	4
7	6	-	1	1
$8 = 2^3$	7	3	1	6
$9 = 3^2$	8	4	1	2
$10 = 2.5$	9	5	2	4
11	10	-	1	1
$12 = 2^2.3$	11	4	3	18
13	12	-	1	1
$14 = 2.7$	13	7	2	4
$15 = 3.5$	14	6	2	4
$16 = 2^4$	15	4	1	24
17	16	-	1	1
$18 = 2.3^2$	17	5	3	8
19	18	-	1	1
$20 = 2^2.5$	19	6	3	18
$21 = 3.7$	20	8	2	2
$22 = 2.11$	21	11	2	2
23	22	-	1	1
$24 = 2^3.3$	23	5	4	96
$25 = 5^2$	24	8	1	2
$26 = 2.13$	25	13	2	4
$27 = 3^3$	26	6	1	6
$28 = 2^2.7$	27	8	3	18
29	28	-	1	1
$30 = 2.3.5$	29	7	6	36
31	30	-	1	1
$32 = 2^5$	31	5	1	120

the 'ultimate' factorisation, and it is this feature of canonicality that induces the minimality of the corresponding Series algorithm. In such cases, the equaliser can be designed as a series cascade of subsystems, whose total number is

$$M = m_1 + m_2 + \ldots + m_k$$

The cascade consists of m_1 subsystems with $p_1 - 1$ taps, m_2 subsystems with $p_2 - 1$ taps, ... and so on, so that the total number of (adaptive) taps is

$$C = m_1(p_1 - 1) + m_2(p_2 - 1) + \ldots + m_k(p_k - 1)$$

In addition, M unity taps are required. We assume that these do not contribute significantly to the complexity.

We now consider some simple cases, with a view to estimating the magnitude of the complexity.

The upper bound is obviously achieved when n is prime: in this case, $C(n) = n - 1$, but this is untypical.

More generally (and equally untypically) if $n = p.q$, where both p and q are prime, then

$$C = \frac{n}{p} + p - 2$$

so that these cases lie on straight lines with gradients $1/p$.

Most cases, however, lie on curves that are logarithmic. For the case $k = 1, p_1 = 2$ we have $n = 2^m$ and the asymptotic order is **exactly** achieved as a lower bound:

$$C = \log_2(n)$$

A similar calculation for all cases with $k = 1$, i.e. $n = p^m$, also gives an **exact** estimate:

$$C = (p - 1)\log_p(n) = \frac{p - 1}{\log_2(p)} \log_2(n) = O(\log_2(n))$$

The case $n = p^m.q$ gives rise to another logarithmic curve:

$$C = \frac{p - 1}{\log_2(p)} \log_2(n) + \left[q - 1 - \frac{\log_2(q)}{\log_2(p)} \right]$$

Finally, we see that the logarithmic cases are typical, by underbounding and overbounding n as follows:

$$p_{\min}^{m_{\min}} \leq n \leq p_{\max}^{m_{\max}}$$

We can show that C is bounded in a similar way:

$$\frac{p_{\min} - 1}{\log_2(p_{\min})} \log_2(n) \leq C \leq \frac{p_{\max} - 1}{\log_2(p_{\max})} \log_2(n)$$

Assuming that the growth of the factor $(p_{\max} - 1)/(\log_2(p_{\max}))$ is so slow as to be negligible, we conjecture that the normal order of C is logarithmic:

$$C = O(\log_2(n))$$

We also present in Fig. 5.21 the variation of complexity C with equaliser order $N = n - 1$, in graphical form for the first 100 cases (stopping at the prime $n = 101$).

Values of N above 10, say, must be regarded for the present as beyond the realm of practical possibility, and these results are presented mainly as conformation of the asymptotic theory outlined above: note the existence of the straight-line cases, and the more typical tendency to cling nearer the logarithmic lower bound.

It is clear, from combinatorial arguments, that the total number of different ways of designing the cascade architecture is the multinomial coefficient:

$$D = M!/(m_1!m_2! \ldots m_k!)$$

Allowing for the permutation of the M subsystems into an arbitrary order, the total number of equivalent configurations is

$$E = M!^2/(m_1!m_2! \ldots m_k!)$$

We have been unable to estimate the magnitude of D and E, despite the availability of material on related problems, such as [16], where it is shown for example that M and k are both of **average** order $\log(\log(n))$.

It is clear that we have here an example of the 'divide-and-conquer' [17] strategy: the complexity reduction is obtained by partitioning the algorithm into subalgorithms in such a way that the total computational workload is diminished. In view of the similarity with the Fast Fourier Transform (FFT) [18], in which a reduction $n^2 \rightarrow n\log_2(n)$ is obtained (also by means of subgroup decompositions of $\mathbb{C}_n$), it is appropriate to designate this approach to equaliser design the 'Fast' Convolution Algorithm. This approach is not, of course, generally applicable to arbitrary signals as in the case of the FFT. On the contrary, it is closely matched to the single-precursor signal structure that we have assumed for the purpose of this study. Thus we may regard these designs as examples of a general principle: if the signal possesses a special structure, it should be possible to exploit this structure by means of an appropriately designed signal processing algorithm. Seen in this light, the designs seem to be far less exotic.

5.5 SUMMARY AND CONCLUSIONS

We referred earlier to an alternative approach using Cyclotomic polynomials. Though less direct, it gives identical results for all the cases considered, and provides a useful corroboration of the designs, all of which (for N composite and greater than three) are novel. We have also examined the use of Galois Field models where appropriate. There are cases, of course, where no Galois Field model can be found (for example, $n = 6$), as the order of such fields must either be prime or some power of a prime. Even so, the additional algebraic structure that accrues from the additivity properties casts further light on those cases where the finite field model applies. In particular it is possible to decompose the entire field into disjoint conjugacy classes indexed by the cyclotomic cosets. This again gives rise to a factorisation of the equaliser transfer function, but in this case in terms of polynomials irreducible over the field concerned [19].

As before, these are not the polynomials required, but the regrouping into

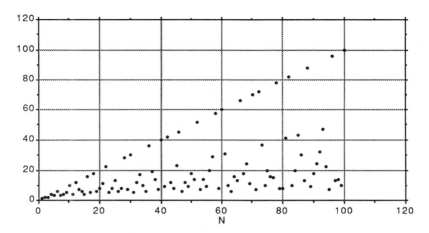

Fig. 5.21 — Variation of complexity (C) with equaliser order (N).

higher-order, necessarily reducible, polynomials is straightforward, and confirms the results of the method described in section 5.2 where applicable. Presumably, there are mathematical techniques for dealing with the *lacunae* in the finite field model for $n \neq p^m$.

On the face of it, these alternative approaches are merely corroborative, and produce no new results. But this situation is likely to change when we consider more complicated ISI models. The design of an equaliser for a second-order signal model, for instance, can easily be shown to be equivalent to finding the factors of the trinomial

$$P(x) = x^{n+1} - x - c$$

where n is the degree of equalisation required. Once found, the factors would have to be grouped in such a way as to minimise the total number of coefficients. There is a possibility that some of the results to be found in [20] might be applicable: certainly factorisations of $P(x)$ exist in terms of lower-order trinomials (provided c is an element of an associated Galois Field). It is not clear whether this is a severe restriction or not, but the decomposition into trinomials appears to be analogous to the case of the series cascade of first-order sections for the single-precursor case: there will be no reduction in complexity without grouping factors.

There is some experimental evidence that equaliser systems currently undergoing development can be adequately designed on the basis of the single-precursor signal model, and further work is in hand to investigate the situation more closely. Even so, there is still a need to extend the theory to second-order and even higher-order models. A better understanding of these models might allow an improvement in the current state of the art by moving beyond the present limits of performance. Furthermore, future systems are likely to employ more complicated multi-level modulation techniques. The narrower bandwidth of these schemes will simplify the distortions to some extent, but longer equalisers are likely to be required, because the sensitivity to ISI increases rapidly with the number of signal states.

We have seen from the preceding chapter that Abstract Algebra (specifically Galois Field Theory) has a part to play in effecting a *rapprochment* between 'mainstream' Signal Processing and Error Control Coding. We suggest that similar mathematical modelling techniques (specifically Group, Field and Lattice Theory) may have a corresponding *role* in the harmonisation of Computing Science, Complexity Theory and Adaptive Equalisation. The approach to equaliser design *via* Abstract Algebra that we have described is rudimentary in that we have only discussed in detail the simplest case — that of the single precursor — for which the algebraic structure is particularly manifest. Work is in hand to determine whether the design theory can be generalised to deal with the multi-precursor signal.

REFERENCES

[1] A. P. Clark, *Advanced Data-transmission Systems*, pp. 175–301, (1977) Pentech Press.

[2] M. T. Dudek, *Series Equaliser*, Priority Patent Application No. 8605518, 6 March (1986).

[3] M. A. Kitching & M. T. Dudek, 'A novel configuration of baseband equaliser for digital radio', *Proc. First European Conf. on Radio Relay Systems*, Munich (4–7 November 1986), pp. 332–338.

[4] L. R. Rabiner & B. Gold, *Theory and Application of Digital Signal Processing*, pp. 28–38, (1975) Prentice-Hall, Englewood Cliffs, NJ.

[5] J. W. Archibold, *Algebra*, pp. 152–157, (1970) Pitman, London.

[6] A. R. Camina & E. A. Whelan, 'Linear groups and permutations', *Research Notes in Mathematics*, Vol. 118, pp. 65–74, (1985) Pitman, London.

[7] J. J. Cannon, 'An introduction to the group theory language CAYLEY', *Proc. LMS Symposium on Computational Group Theory* (Durham 1982).

[8] B. Higson, *Applied Group-theoretic and Matrix Methods*, (1964) Dover, New York.

[9] J. A. Conway, R. T. Curtis, S. P. Norton, R. A. Parker & R. A. Wilson, *Atlas of Finite Groups*, (1985) OUP.

[10] D. E. Rutherford, *Introduction to Lattice Theory*, (1965) Oliver and Boyd, Edinburgh.

[11] A. R. Camina & E. A. Whelan, 'Linear groups and permutations', *Research Notes in Mathematics*, Vol. 118, pp. 99–101, (1985) Pitman, London.

[12] K. V. Lever, E. Patterson, P. C. Stevens & I. M. Wilson, 'Surface acoustic wave matched filters for communication systems', *The Radio and Electronic Engineer*, **46**, 237–246 (1976).

[13] K. V. Lever, E. Patterson, P. C. Stevens & I. M. Wilson, 'Surface-acoustic-wave matched filters for multi-frequency shift keyed communication systems', *Proc. IEE*, **125**, 770–774 (1976).

[14] W. W. Peterson & E. J. Wheldon Jr, *Error Correcting Codes*, pp. 472–492, (1980) 2nd edn, MIT Press.

[15] D. F. Stanat & D. F. McAllister, *Discrete mathematics in Computer Science*, pp. 232–236, (1977) Prentice-Hall, Englewood Cliffs, NJ.

[16] G. H. Hardy & E. M. Wright, '*An Introduction to the Theory of Numbers*, pp. 260–272, (1975) Fourth Edition, OUP.

[17] D. F. Stanat & D. F. McAllister, *Discrete Mathematics in Computer Science*, pp. 248–256, (1977) Prentice-Hall, Englewood Cliffs, NJ.

[18] L. R. Rabiner & B. Gold, *Theory and Application of Digital Signal Processing*, pp. 371–379, (1975) Prentice-Hall, Englewood Cliffs, NJ.

[19] F. J. MacWilliams & N. J. A. Sloane, *The Theory of Error-correcting Codes*, pp. 93–109, (1977) North-Holland, Amsterdam.

[20] R. Lidl & H. Niederreiter, 'Finite fields, *Encyclopedia of Mathematics* **20,** p. 128, (1983) Addison-Wesley, Reading, MA.

Part 2

Communication security

There is an increasing amount of transference of information concerning personal matters which either the sender or the receiver would prefer to be kept secret or at least inaccessible to some third parties. The privacy of a telephone conversation and the reliability of a cash card withdrawal are two examples. In these circumstances the issue of communication security is extremely important and is the central theme of Part 2 of this volume. In practice there is often a compromise to be struck between the level of security offered by a system and the amount of time and expense committed to the communication process.

In the first chapter of this part, Piper identifies the need for security and some areas of risk. He also gives a wide-ranging review of techniques that have been, and are being, developed in this field. As far as privacy is concerned, data encryption is important and access to random numbers is fundamental in several ways for the development of cryptosystems. A description of some new, reliable algorithms for their generation is given in the second chapter. In the next, Bromfield gives an overview of security problems in speech communication and describes methods available for their solution. Finally, Davies develops some elegant results and ideas which bring out close analogies between cryptography and error correction.

It is clear from the contributions that this is a relatively new and expanding area of application for mathematics in which much important and interesting work is under way.

6

Information security

Fred Piper
Royal Holloway and Bedford New College, University of London, UK

6.1 INTRODUCTION

The banks now regularly perform high-value electronic transfers involving billions of pounds. This is just one reason why Information Security has been one of the growth industries of the 1980s. The Big Bang and the increase in the popularity of electronic mail are two others. Furthermore, many countries now have data-privacy legislation which imposes a legal obligation to protect certain information.

There are two different types of attack which might be launched against a communications system or data network: passive and active. In a passive attack the aim is merely to gain unauthorised knowledge of confidential information. Passive attacks may be thwarted either by denying the would-be attacker access to the information or by encrypting it so that, even if he does gain access, it is meaningless to him. The aim of an active attack is different. Here the attacker is trying to alter information. It is often difficult to prevent the alterations of messages, and, consequently, these attacks are usually combated by taking steps to ensure that any alteration will be detected by the receiver. It is perhaps worth noting that it is becoming easier to gain access to data files on hosts and to intercept transmissions between links. Indeed for a number of popular media, e.g. satellite links and radio telephones, accidental interception is a distinct possibility.

In this chapter we look at the need for security and point out a number of recognised risk areas. When discussing the techniques for providing security, we concentrate on three areas:

> Access Control;
> Message Authentication;
> Privacy (i.e. message encryption).

In each case we concentrate on the security requirements, the techniques available, the relevant international Standards and recent developments.

Two results of recent technological advances and the increase in public awareness of the need for security are cheaper and improved data security. Coupled with them, however, is the disadvantage that security is much more in the 'public eye' and that would-be attackers are much better informed about current techniques. The dramatic increase in the power of home computers enables them to perform long, detailed cryptanalytic attacks. Furthermore, fairly sophisticated bugging devices are now widely available.

6.2　THE RISKS

There are so many areas where a communications system and/or data network might be attacked that it is impossible to list them all. We will merely highlight a few of the obvious ones. Any hardware might be either physically stolen and/or tampered with. Such hardware might include processors, terminals or even purpose-built security modules. Software might contain trapdoors, logic bombs or trojan horses. Secure networks are likely to contain databases of highly sensitive information. In fact, they often contain information, such as encryption keys, which is crucial to the overall security of the system. A major risk is that someone will obtain illegal entry to such databases and be able to read the relevant information. A further obvious risk for any system is that certain people are likely to have access to secret information or to be needed to perform crucial tasks. All personnel must be trustworthy, in terms of both honesty and competence.

If information is to be transmitted from one location to another, then that communication is likely to be subjected to either passive or active attacks. If the information is highly secret, e.g. military or government communications, then the risk of a passive attack is of great concern. However, there are an increasing number of applications where the information being transmitted is in no sense 'secret' but where its integrity must be guaranteed. The obvious examples are the teleprocessing systems for handling financial/commercial transactions. Such systems range from those transferring billions of pounds between major banks to those where customers use plastic cards to pay retailers. A common need for all such systems is to achieve integrity so that all parties can have confidence in the outcome of the transactions performed on their behalf. The enormous amounts involved in some of these transactions guarantee that the criminal world will attempt to profit from unauthorised and/or altered transactions. There is therefore an obvious need for the designers of these systems to appreciate the type of active attacks which are likely to be launched against them and to introduce appropriate security countermeasures.

The following are (necessarily incomplete!) typical examples of attacks on systems.

Tapping of communication; reading messages and addresses.
Replaying messages.
Diverting messages.
Overloading system.
Modification to local network station logic.
Bugging of terminal.
Bugging of computer bus, e.g. store data bus.
Copying of disk and stealing copy.
Observing entry of password.

6.3 STANDARDS

The last ten years have seen the formulation of numerous data-security Standards eminating from a number of different sources. These Standards range from an enciphering algorithm (The Data Encryption Standard or DES) and modes of operation of DES to procedures for key management and techniques for message authentication. Apart, possibly, from an actual algorithm there is an obvious need for certain internationally accepted Standards. The mere fact that the major plastic credit cards are a standard size has meant that dealers can use one 'machine' to handle transactions for different card issuers — an obvious financial saving.

In addition to being crucial for interoperability, the existence of Standards is clearly beneficial for organisations who need security but have no real expertise in the area. The knowledge that their system conforms to an internationally accepted Standard will increase their confidence in it. This may not, however, be true of encryption algorithms, because here many users like to feel that their algorithm is unique to them. In this situation, knowledge that someone else is using 'their' algorithm often reduces confidence rather than increasing it.

For data security, the two major standard 'authorities' are probably the American National Standards Institute (ANSI) and the International Organization for Standardization (ISO). There are, however, many others. Some are national, whereas others have restricted interests, e.g. the International Telephone and Telegraph Consultative Committee (CCITT).

6.4 PRIVACY

The fact that we are discussing privacy before both access control and message authentication should not be interpreted as implying that it is more important. Our reason is that a number of message authentication techniques and some recent access control systems employ the major technique for ensuring privacy, i.e. encryption.

The basic idea behind any data-encryption algorithm is to define functions f_k, which transform messages into cryptograms, i.e. disguised forms of the original messages, under the control of secret keys. Thus, if we let M denote the set of all possible messages, and C denote the set of all possible cryptograms, then we are defining a family of functions $\{f_k : k \in K\}$, where K is the set of all possible keys, and $f_k(m) \in C$ for every $m \in M$. In order that the encryption function be usable, decryption must always be possible, which means that every f_k must be one-to-one (i.e. $f_k(m) = f_k(m')$ implies $m = m'$).

This rather abstract notion of data encryption is not necessarily a good guide to classifying the techniques actually used in cryptographic applications. In general, the idea of a special function being applied to the entire message simultaneously in order to obtain the cryptogram is rarely, if ever, used. In practice, most encryption techniques involve dividing a message into a number of small parts (of fixed size), and encrypting each part separately, if not independently. This greatly simplifies the task of encrypting messages, particularly as messages are usually of varying lengths.

We will not discuss encryption algorithms in general. However, we must divide them into two rather important classes and say a little about each. The division is into conventional (or symmetric) and public key (or asymmetric key) systems. The fundamental difference is that in an asymmetric system, encryption and decryption are sufficiently different processes as to require different keys, and knowledge of an enciphering (or deciphering) key is not in practice sufficient to be able to deduce the corresponding deciphering (or enciphering) key.

The central principle on which public key systems are based is the distinction between encrypting keys and decrypting keys. Instead of having a single set K of keys k used for both decrypting and encrypting, we now have a pair of sets K_E and K_D of encrypting and decrypting keys respectively, and a one-to-one correspondence between the elements of the two sets. In addition we have two families of functions F and G, where F consists of encrypting functions and G of decrypting functions, where the elements $f_e \in F$ are indexed by $e \in K_E$, and the elements $g_d \in G$ are indexed by $d \in K_D$.

The property that these sets must satisfy is that for any encrypting key $e \in K_E$, if $d \in K_D$ is its corresponding decrypting key, then $g_d(f_e(m)) = m$ for every message m. The way in which such systems are used is that a user generates a matching pair of keys (e,d), where $e \in K_E$ and $d \in K_D$ and keeps d secret whilst publishing e. The encrypting key e is then called the public key.

In order for such a system to be effective, knowledge of an encrypting key e must not in practice enable anyone to compute the matching decrytping key d (although, of course, it will always be possible to do this in theory). To make the scheme useful in practice, the encryption scheme must include a convenient means of generating pairs of encrypting and decrypting keys, whilst retaining this property.

The most well known and widely used conventional algorithm is the Data Encryption Standard (DES) algorithm,. This was oroginally specified as a US National Standard in 1977 and has become a *de facto* international standard.

DES is a block cipher. It has a 56-bit key and operates on 64-bit blocks. Ever since its introduction it has been repeatedly criticised, particularly for the relatively small size of the key used. People have claimed that it would be comparatively cheap or easy to build a machine which would perform an exhaustive key search (i.e.try all possible keys) in a relatively short time. However, there has been no obvious sign of the existence of such a machine, and it is still widely believed to be a more-than-adequate technique for commercial applications.

When assessing the security level of a communications system it is usually considered unwise to rely on the algorithm being secret. This means that the security of the cryptogram relies on the secrecy of the deciphering key. For conventional key systems this is, essentially, the same as the secrecy of the enciphering key, and maintaining the secrecy of these keys is one of the major 'headaches' when designing a secure system,. Key management is one of the most difficult aspects of designing a system. Keys have to be generated, distributed, stored and, when no longer needed, deleted. Each operation must be performed securely because if an attacker obtains the keys being used then, no matter how strong the algorithm may be, he will have the same knowledge as the genuine receiver and will be able to deduce the message from the cryptogram.

If we restrict our attention to key distribution then we immediately come across an interesting problem. How can we secure the keys during the distribution? One solution is to use an alternative communications link, e.g. couriers. For small systems this may be feasible. (However, for the actual messages, this use of a courier would be both too slow and too expensive.) There are, however, many large networks or even small ones, e.g. certain satellite links, where there may be no alternative to transmitting the keys across the same link. In such instances the keys need to be encrypted and it is therefore necessary to introduce a key hierarchy including, for instance, data-encrypting keys and key-encrypting keys. In most key hierarchies, the top-level keys need to be kept in physically secure locations. Any keys which are kept in physically insecure terminals (e.g. point-of-sale terminals) are likely to need frequent changing.

Some of the more obvious problems associated with key management for conventional systems do not arise if a public key system is used. For a public key system the encryption keys are public and so, obviously, do not need to be transmitted securely. In fact, one possible implementation of a public key system might involve users generating their own keys and then sending the public key to a directory which may be consulted by everyone. At first sight this appears to eliminate the key distribution problem. However, other problems arise. One is the need to ensure that the public key stored in the directory is the one which is actually sent by the user. There are obvious advantages for a criminal to have his own public key registered as that of someone else. Another potential problem associated with the use of a public key system is that the receiver has no natural way of identifying the sender. There are many situations where each communicator needs to be absolutely certain of the identity of the other. (The transfer of money is one such example.) There are, fortunately, a number of protocols for overcoming this latter problem, and a number of systems currently use a public key algorithm for securing the keys of a conventional system.

Finding suitable algorithms for public key systems is proving difficult. At present the only one which appears likely to be widely used is the celebrated RSA algorithm. Here the public key is a pair of suitably chosen h and n and, for a message block m (where m is an integer less than n), the encrypted form of m is $m^h \pmod{n}$. The security of the system depends upon the interceptor being able to factor n, and consequently n must be large. (To provide a security level comparable to that of DES, n should be about 400 bits. A commonly recommended size for n is 512 bits.) Since encryption involves so much computation it is a slow process and this is one of the reasons why RSA is rarely (if ever) used to encipher data. There are numerous rumours, claims, etc., about RSA chips either existing or being 'about to exist'.

However, none seems to be claiming encryption and decryption rates of more than about 64 kbits/s for a 512-bit modulus.

6.5   MESSAGE AUTHENTICATION

There are a number of different requirements for message authentication. They include:

Detection of message alteration.
Detection of message deletion.
Detection of message replay.
Transparent to user.

Minimal system impact.
Full audit facility.

The need for protection against message alteration and deletion is probably obvious. An instance where message replay might be disastrous is provided by the withdrawal of cash at an ATM. Immediately prior to the customer receiving his cash from the ATM, the ATM receives an instruction from the bank's computer authorising that payment. If a criminal were able to repeatedly replay that last instruction, then he would be able to 'milk' the machine of all its cash.

The need for a full audit trail is crucial. If an attack takes place, whether successful or not, it is vital that the network manager can detect where, when and by whom. Preferably the detection should take place before the attack has had a chance to suceed, but we will say more about that when we discuss access control.

The most common way of providing message authentication is to add a cryptographic check sum (called a MAC) to the end of the message and to transmit it with the message. This check sum will depend upon a (secret) key, so that anyone who changes part of the message will not be able to make the appropriate alteration to the MAC. It should also be such that a change in any message bit is likely to affect its value. Of course the addition of the MAC adds message extension and it is clearly desirable to minimise its length. It must, however, be long enough that the chances of it being correct by chance are minimal. A commonly recommended length for a MAC is 32 bits. (In fact there is an ANSI Standard for using DES to produce a 32-bit MAC.) It is important to note that message authentication does not offer privacy. However, there is no reason at all why the MAC cannot be a function of some secret data (which are not transmitted) in addition to the message.

6.6   ACCESS CONTROL

There are at least two different aspects of access control. The first is to ensure that only legitimate users gain access to the system. This is achieved by requiring each user to prove his identity prior to logging in. The other aspect is to define the set of functions, data-files, etc., that bona-fide users are allowed to access once logged in.

6.6.1   User verification

If we wish to install access control on a computer system then we need a way of recognising the genuine users. Thus we need a process whereby if a user, possibly at a terminal which is remote from the computer, enters a user identity then the system can verify that the user has the claimed identity. It is worth noting that this process, called user verification, only involves checking the validity of a particular individual's claim to a specific identity. It is therefore, different and easier than the problem of user identification. As an illustrative example we can consider the problems faced by the police when trying to determine the 'owner' of fingerprints found at a crime. If they have a suspect then it is straightforward to check his fingerprints with those found at the crime. This is user verification. If, on the other hand, they have no suspect then they have the more difficult task of trying to determine the culprit from the prints. This is user identification.

Most attacks levelled against networks are instances in which an unauthorised person manages to find a way through the log-in system. In practice this type of attack usually involves stealing or guessing a user identity that is recognised by the system and then stealing or guessing a password which 'belongs to' that identity. The reason for the success of these attacks is that over 95% of existing computer networks use access control systems which merely involve entering a user identity together with the appropriate password. Since there is no other protection, it is hardly surprising that hackers succeed. Furthermore, in almost all of these systems the entry of an incorrect password is not even recorded. Thus there is absolutely nothing to prevent, or even discourage, an attacker systematically trying all passwords until the correct one is found.

In this section we will assess some of the user identification techniques which are currently available and follow this by looking at an application for a particular user verification technique.

Passwords

Password schemes are notorious for the low level of security that they offer. One of the major reasons for this was admirably expressed by Alan Vane in an article called 'Password to theft' which appeared in the *Financial Times* on 6 September 1986. He wrote: 'Hackers get inside systems because they get hold of passwords, and humans, being human, are careless about password control. In other words, the thieves get hold of the key to the cash box, left carelessly unguarded, rather than pick a lock with superhuman skill' There are essentially two ways to obtain a user's password. One is to capture it; the second is to guess it.

The simplest technique is to ask the user and, surprising though it may seem, this has a reasonably high success rate. Another simple method is to watch the user enter his password and to record it.

When the plastic cards issued by banks are used to gain access to their computer networks, the customer's Personal Identification Number (PIN) acts as his password. Customers are always advised to keep their PINs secret.

In the USA recently, a team of fraudsters attempted to capture a number of account numbers (equivalent to user identity) and PIN pairs. To do this they set themselves up as a mail order company and advertised some goods that could be bought on a debit card simply by phoning in. (Clearly such transactions require the disclosure of account numbers over the telephone.) As unsuspecting customers phoned in, they were, quite correctly, asked for their debit card numbers so that their accounts could be debited to the value of the goods. They were, however, also asked for their PINs. Although the PIN is, of course, not needed for this type of transaction it is alleged that most customers gave their PINs freely. Those that did question the requirement for their PINs were told that the PINs were needed to make the debits from their accounts. It is reported that few customers failed to give the information requested. The fraudsters, meanwhile, were manufacturing cards which, together with the now known PINs, they then used to withdraw cash on those customers' accounts from the Automatic Telling Machines (ATMs) which dispense cash for customers who enter a card and then punch in the correct PINs.

As far as guessing is concerned, the most popular passwords in Britain are 'pass' and 'Fred', whereas the Americans focus on love and sex. According to the Hogg Robinson report, hackers guessing at passwords can expect a success rate as high as

20% within surprisingly few trials. In order to attain more security it is necessary to install a user verification system that makes it significantly harder for the user to be impersonated.

Identity verification is based classically on something known to the user, something possessed by the user or some physical characteristic of the user. 'Something known' usually implies a password or, in the case of systems operated by banks, the 'PIN'. 'Something possessed' may be a token, for example a plastic card or key. Physical characteristics used for identity verification are often called 'biometrics'; examples include voice and signature verification. It is generally considered that they provide security superior to that offered by PINs, but there are other disadvantages which discourage their widespread introduction, not least the degree of imprecision that is inherent in many of the techniques proposed.

There is no doubt that the most common technique for user verification is that which uses the plastic card; there are millions of such cards in use throughout the world today, most of which carry a magnetic stripe. Strictly speaking, the card only allows the identity to be claimed and does nothing in itself to help the verification of that identity. To prevent a lost card being used by someone other than the authorised user, it is customary to require that the user offer a memorised PIN in conjunction with the card; a derived function of the PIN is often stored on the card. Human memory being what it is, the 'memorised' PIN is often written down, even on the card itself; the problem is compounded by many users possessing several cards, each with its own different PIN.

The magnetic stripe card is itself a source of insecurity for systems in which it is used. The problem is that the card is too easily counterfeited. The magnetic stripe can easily be copied unless special measures are taken to make this difficult, the embossing can easily be altered or simulated, and the general appearance of the cards can be imitated in a way sufficiently sophisticated to deceive human counter clerks. Various security measures have been added to the cards in recent times to make these falsification methods less effective; the hologram on the face of the card and the use of 'watermark' magnetic tape are examples of such measures.

Because of the shortcomings of the magnetic stripe as a secure means of recording user parameters, a great deal of attention has been paid in recent years to alternative card technologies. In particular we have the intelligent or 'smart' card, where integrated circuits (one or more) are embedded within the plastic; surface contacts are usually provided as part of a communication interface, though other means of communication are also being considered. It is well known that experiments have been carried out, particularly in France, to assess the performance of 'smart' cards. The outcome of these experiments has not been widely publicised, but the French are showing their confidence in these devices by launching new applications, with millions of cards in use.

It seems that counterfeiting of the smart card presents the criminal with much greater problems than are met with the magnetic stripe card. However, the problem of the PIN as a means of confirming claimed identity remains. Presentation of a PIN associated with the smart card requires that it be entered on a keyboard associated with the terminal. Because the field of application of the smart card is bound to include transactions such as point of sale, the insecurity of the average point-of-sale terminal is significant. Physical protection of such terminals is bound to be much less than that normally found in automatic teller machines. Therefore we may expect that some point-of-sale terminals will be bugged, with the object of collecting personal account information and PINs. Exploitation of this knowledge may not necessarily involve fabrication of false cards; stealing of a smart card whose PIN has been discovered is a much simpler means.

Biometric techniques

PINs and passwords are both pieces of information which are given to the user in order to help prove his identity. Ideally he should be the only person to know this information and should remember it so that it is not stored anywhere. In the systems described so far, knowledge of this information, often together with possession of a token, is taken as proof of identity. This clearly has a number of drawbacks including, for instance, the one that an attacker might steal the token and obtain, or even guess, the PIN. We have already seen many ways in which this might happen, and if it does then, unless the loss of the token or compromise of the PIN has been reported, the system will inevitably identify the attacker as the genuine user. Clearly there would be some advantages if the identification process involved something which was more natural to the user, for example, his signature or some physiological characteristic which identifies him uniquely. Schemes which attempt to use physiological characteristics for user verification are called biometric. There is a wide range of biometric techniques, which are currently either in use or being considered for use. They are based on such varying properties as signatures, fingerprints, retina scans, palm scans, vein scans, bumps on the head, and so on. Clearly, one of the problems associated with introducing such techniques for user verification is user acceptance. Most people are likely to feel that the idea of having a laser directed at their retina every time they want to make a financial transaction is totally unacceptable!

A more important, theoretic, problem concerns the trade-off on such systems between what are called the type I and type II errors. In order to explain this, let us look more carefully at the biometric technique.

The basic idea behind all biometric systems is to measure some physiological characteristic of the person, his fingerprint, say. In order to achieve this, a sophisticated digital signal processing technique will then be employed to 'read' the fingerprint, and a number of parameters will then be stored which (ideally) uniquely characterise that fingerprint. Unfortunately, in practice, each time the fingerprint is read, the value of the parameters will be slightly different. This may be due to any one of a number of reasons including, for example, slightly different positioning of the finger, dirt on the finger, scratches and body temperature. Thus in practice the idea will be to store a 'template' of the fingerprint and, every time the user's identity is to be verified, the fingerprint will be read and that reading will be compared against the template. A decision will then be taken as to whether the reading and template match sufficiently well to give a positive verification. The template is normally formed from an average of a number of readings and, with the more sophisticated systems, might even be updated every time a new positive reading is taken. (Note that this latter technique helps take account of characteristics that change with time.)

The decision as to whether a positive verification has taken place is achieved by means of a measure which indicates how near the current reading is to the template. Thus a threshold is set and the decision is made as follows: if the measure is less than the threshold value, this is the person claimed; if the measure is higher than the threshold, this is not the person claimed.

A type I error is when a legitimate user is rejected. The likelihood of a type I error is often called the 'insult' rate since it measures the percentage of legitimate users that will be rejected.

A type II error occurs when an imposter is accepted. The likelihood of a type II error is thus, in some sense, a measure of the security of the system, at least in terms of the resistance to this kind of attack.

Clearly, by suitable choice of the threshold, one can influence the likelihood of errors of either type. For instance, if we feel the current insult rate is too high, we can increase the threshold until we have a more acceptable rate. Unfortunately in so doing we inevitably increase the possibility of a type II error, i.e. we are decreasing the security of the system. Clearly, systems will only be used when the security level is acceptable to the bank and the insult rate is acceptable to the user. It is very difficult to find the right balance between these conflicting requirements. The balance which is acceptable varies dramatically with the application.

Suppose, for instance, we are considering a biometric system for access control for the security manager to his user database. Then it is absolutely crucial that only he can gain access, and consequently security is the only real requirement. Although he may find it frustrating to be rejected type I errors are not important and he will, accept a (reasonably) high insult range in exchange for no type II errors.

In contrast we now consider the situation for an electronic funds transfer system at the point of sale. Suppose we have a system with a type I error of 0.1%. Although this may sound low it means that about 1 in every 1000 genuine cardholders arriving at the point of sale, perhaps in a supermarket with a trolley load of goods, will be rejected (in full view of everyone!) as being a fraudster. Such customers will be embarrassed and annoyed. Furthermore, they will almost certainly change their bank.

Dynamic passwords
The idea of a dynamic password is relatively new. However, its introduction was, we believe, one of the most important recent breakthroughs in the area of user verification.

The basic idea is that, as the name suggests, a user's password should change frequently and that this change should be influenced by some secret information and/ or an intelligent device which is unique to that user. This is usually achieved by the following type of scenario.

The computer that is trying to verify the user's identity selects a 'challenge' and issues it to the user. The user then computes his 'response' to the challenge in such a way that the response depends on:

(1) the challenge that was issued to him;
(2) a token which he possesses;
(3) his (secret) information.

As an example, the user might, at his remote terminal, use a token which, for reasons which will soon be apparent, we have called a 'generator'. As soon as the user enters his identity at the terminal, a challenge is displayed on the terminal screen. This challenge is, typically, as short numeric string (probably seven or eight digits). The user then activates his generator by entering his secret information via a keypad. This information will probably be a string of numbers, i.e. a PIN, and will consist of four to eight digits. Assuming that the PIN is correct, the user will then read the challenge from the terminal screen and enter it into the generator. Within the generator a function will then calculate a response whose value is a fixed function of the challenge. However, each generator will have a different function, so the response to any challenge depends on the generator, and therefore, on the user. Thus this response will be the user's 'one-time' password to establish his identity. On the next occasion that this user needs to establish his identity, a different challenge will be issued by the system with the result that the user's password will change. But, of course, the system will be able to recognise his new password as 'correct' and will be able to verify the user's identity.

The Access Controller might be a peripheral device to the host. However, it might also be embedded in the host software as part of the log-in program or implemented as a front end to the host. The Access Controller is responsible for issuing the challenges, verifying the responses, keeping an audit trail and maintaining the user database.

It is clearly necessary for different generators to produce distinct responses to the same challenge. One way of achieving this is to use a cryptographic function and issue each generator with a unique key. In this instance the generator requires two inputs: the PIN, known only to the generator user, and the challenge. It then produces a single output, namely the response. The PIN essentially provides access to the device, thus protecting the user against loss of his generator. The device itself is programmed, prior to its issue to the user, with a cryptographic key. Normally that key will be unique to that user, and the host controller will maintain a database of keys against user identities. When the user enters the correct PIN into his device, the appropriate key will be selected. If he enters the wrong PIN, an incorrect key will be used. The key selected plus the challenge are then processed through the data encryption algorithm, possibly DES, to produce a response. It should be noted that if the generator is sensibly designed, then no matter what PIN or challenge is entered, the device will always produce a response. This ensures that the only way that an unauthorised user can make use of a stolen generator is by going through the system. In other words, the generator itself will give him no clue as to whether or not he has entered the correct PIN. He can only discover this by sending his response to the controller and having it either accepted or rejected. It is now fairly straightforward to prevent an attacker systematically searching for the PIN. As soon as he has one false attempt, an invalid response is recorded at the controller. It is then possible to program the controller to raise an alarm or remove a user from the system following more than a designated number of invalid responses in a given time.

6.6.2   Permissioning
Even when a user has been allowed to log in, it may be necessary to limit his activities. Certain functions, such as modifying the access control system itself, must be strictly limited.

The term 'Security Reference Monitor' (SRM) is used to define the software kernel which performs this function. The integrity of the SRM must be ensured, and so it is often known as the 'Trusted Computing Base' or TCB. For it to be described as 'trusted', it will have been assessed and validated by some independent agency.

A scheme for limiting access, which is used by a number of TCBs, is the 'Bell and LaPadula Model'. Here, a system of hierarchical classification levels and/or non-hierarchical categories is applied to each data object and to each user. Two rules are defined:

> The Simple Security Property, which states that a user's classification must dominate an object's to observe it. That is, his classification level must be greater than or equal to and his classification categories must be a superset of or equal to the object's. In this way, a secret user cannot read a top secret file.

> The * Security Property, which states that a user's classification must be dominated by the object's in order to alter it. Thus a secret user cannot write to an unclassified file. At first sight, this rule may seem unnecessarily restrictive. Its purpose, though, is to guarantee that a secret user can never give secret information away, either accidentally or intentionally. Once information has become classified, it cannot be declassified.

An alternative means of providing the same type of restrictions is to introduce multilevel encryption schemes where certain keys are able to decrypt cryptograms enciphered using 'lower-order' keys, etc. There is a lot of research activity in this area.

BIBLIOGRAPHY

The following is a list of some recent books on cryptography.

[1] H. Beker & F. Piper, *Cipher Systems*, (1985) van Nostrand-Reinhold, Wokingham, UK.

[2] H. Beker & F. Piper, *Secure Speech Communications*, (1986) Academic Press, Orlando, Florida, USA.

[3] D. W. Davies & W. L. Price, *Security for Computer Networks*, (1984) Wiley, Chichester.

[4] D. Denning, *Cryptography and Data Security*, (1982) Addison-Wesley, Wokingham, UK.

[5] A. G. Konheim, *Cryptography: a Primer*, (1981) Wiley–Interscience, Chichester.

[6] C. H. Meyer & S. H. Matyas, *Cryptography: a New Dimension in Computer Data Security — a Guide for the Design and Implementation of Secure Systems*, (1982) Wiley–Interscience, Chichester

[7] F. Kranakis, *Primality and Cryptography*, (1986) Wiley & Taubner, Chichester.

[8] R. A. Rueppal, *Analysis and Design of Stream Ciphers*, (1987) Springer-Verlag, New York.

7

Ultra-uniform pseudorandom number algorithms for cryptokey generation

G. J. Janacek and **K. V. Lever**
University of East Anglia, University Plain, Norwich, NR4 7TJ, UK

7.1 INTRODUCTION

Random numbers find application in several ways in the design of secure communication systems: a compact overview of the subject is provided in [1]. Random numbers are used to provide signal permutations for random time-multiplexing in digital cryptosystems [2], for frequency-band transposition in analogue cryptosystems [3–5], for frequency-hopping in spread spectrum systems [6], and for generating keys in public-key cryptosystems [7,8].

The celebrated RSA algorithm [9] requires the generation of two large prime numbers of typically 256 digits, the product of which forms the 512-digit public-key sequence. The theory of the RSA algorithm and state-of-the-art implementation are reviewed in [10]. Because such systems often require changes in the key, it is necessary to be able to generate new keys quickly and randomly. The approach taken is to generate numbers at random and test them for primality. The choice of an effective primality-testing strategy is discussed in [11] and algorithm structures are described in [12].

We now consider some of the ways of generating sequences of random numbers.

7.2 GENERATORS

Broadly, there are three ways to generate random numbers:

(a) genuinely random devices, such as a noise diode;
(b) maximal-length binary shift-register m-sequences;
(c) linear congruential methods.

The first of these is the easiest to dismiss, even though it is the only source of 'truly' random numbers. The grounds are the low speed and the expense and difficulty of checking that the generator has a satisfactory performance. Even experts have difficulties, as can be seen from the experiences of Kendall and Babington-Smith [13]. For this reason, **pseudo**random numbers generated by some form of finite-state-machine are preferred.

Digital feedback shift-registers are fast and cheap and produce good sequences of pseudorandom bits, provided that subsequences are further randomised by one or more stages of pseudorandom permutation. A description of this technique can be found in [14].

The third approach is to generate a sequence of integers $\{X_t\}$ in the range $[0, m-1]$ by means of a **linear congruential recurrence relation** whose most general form is

$$X_t = (aX_{t-1} + c) \bmod m$$

'seeded' by the initial condition X_0.

The method first appeared in the late 1940s and has subsequently been subjected to comprehensive study: the state of knowledge up to roughly 1982 is summarised in [15]. To obtain satisfactory sequences, certain ground rules must be followed in choosing the parameters a, c and m:

The sequence has period m iff:

(1) c is relatively prime to m;
(2) $b = a - 1$ is a multiple of every prime p dividing m;
(3) b is a multiple of 4 if m is a multiple of 4.

As far as randomness is concerned, c is irrelevant: it merely provides a set of affine transformations of the fundamental sequence. We shall concentrate on the simplified version:

$$X_t = aX_{t-1} \bmod m$$

·In this case we have maximal period when:

(1) X_0 is relatively prime to m;
(2) a is a primitive element mod m;
(3) b is a multiple of each prime dividing m.

Such generators have been exposed to a large number of empirical tests in which the statistical properties of the output sequence are checked for various desirable features, such as runs, gaps, particular combinations, and so on. But as Knuth points out [16], statistical tests are not stringent enough, and it is far better to subject the **structure** of the algorithm to a **theoretical** test.

The **Spectral Test** devised by Coveyou and MacPherson [17] was originally conceived as a comparison of the discrete Fourier transform of a general k-tuple of samples from the generator with the characteristic function that would be expected for a set of genuinely random points uniformly distributed in k-space. Clearly, we are observing the distribution of k-tuples over a lattice of m^k points, and many sites in the lattice are bound to be vacant, since the maximum number of distinct values taken by the sequence is only m. The Spectral Test uses a criterion which measures the degree of clustering of the occupied lattice sites. The figure of merit is the smallest frequency of wavenumber, v_k, in the k-dimensional Fourier analysis of the generator output: if this is small, the k-dimensional, distribution will exhibit conspicuous granularity, and the generator will have poor performance. Marsaglia [18] noted that the k-tuples all fall on a set of hyperplanes, and interpreted $1/v_k$ as the distance between hyper-planes. Thus the larger v_k, the smaller the granularity and the better the generator. Knuth suggests that a generator can only be considered reliable if the following design criterion is satisfied:

$$v_k \geqslant 2^{30/k}$$

Table 1 in [19] lists (the integer values of) v_k^2 for a number of well-known generators. Some of these are very bad, and this emphasises the fact that designing linear congruential pseudorandom number generators has pitfalls for the unwary.

7.3 THE WICHMANN–HILL GENERATOR

Another problem with linear congruential generators is that m needs to be large to provide an acceptably long period, and this will imply difficulty in implementing the scheme on a short wordlength machine. Even mainframe computers are restricted in the choice of m, and the problem becomes acute for mini- or micro-compuers. Naturally one can circumvent the problem by using the appropriate assembly language to concatenate bytes to obtain the required wordlength, but this implies a loss of portability, and the generator becomes prone to misdesign from programming errors.

These problems have been solved by Wichmann and Hill [20–22], who suggested that the following algorithm is both portable and efficient:

$$X_{t\times1} = 171X_t \bmod 30269$$
$$Y_{t\times1} = 172Y_t \bmod 30307$$
$$Z_{t\times1} = 170Z_t \bmod 30323$$

These three individually mediocre linear congruential generators are then norma-lised and combined to form a very uniformly distributed pseudorandom variable in the interval [0,1]:

$$U_{t+1} = (X_{t+1}/30269 + Y_{t+1}/30307 + Z_{t+1}/30323) \bmod 1$$

with period l.c.m.$(30268,30306,30322) \cong 7.0 \times 10^{12}$.

The algorithm can be implemented as it stands on a 16-bit machine in four lines of code, and with only minor modifications in seven lines on an 8-bit machine: it is **extremely portable.** Extensive empirical tests were performed showing that the generator is **exceptionally reliable.**

We were impressed with the quality of this generator and have used it in our work on the design of random processes having specified non-uniform probability density and specified non-uniform power spectral density [23]. Our only reservation was that the generator had not been tested theoretically. In particular, the Spectral Test can only be applied to a **single** linear congruential generator — not to a combination of several. We overcame this problem by showing that the Wichmann–Hill generator is in fact equivalent to a single linear congruential generator.

Consider the following generalisation of the Wichmann–Hill scheme:

$$X_{t+1} = AX_t \bmod p$$
$$Y_{t+1} = bY_t \bmod q$$
$$Z_{t+1} = cZ_t \bmod r$$
$$U_{t+1} = (X_{t+1}/p + Y_{t+1}/q + Z_{t+1}/r) \bmod 1$$

where the moduli p, q and r are distinct primes.

We can show [24], by means of the Chinese Remainder Theorem [25], that the above scheme is equivalent to the following linear congruential generator:

$$W_{t+1} = (qrM_1a + prM_2b + pqM_3c)W_t \bmod M$$

where the modulus M is given by

$$M = pqr$$

and the 'magic' numbers M_1, M_2 and M_3 are easily found from the following relationships by means of the extended Euclidean algorithm:

$$qrM_1 = 1 \bmod p$$
$$prM_2 = 1 \bmod q$$
$$pqM_3 = 1 \bmod r$$

The period is l.c.m.$[(p-1)(q-1)(r-1)]$.

For the actual choice of parameter values made by Wichmann and Hill, we have

$$W_{t+1} = 16{,}555{,}425{,}264{,}690W_t \bmod 27{,}817{,}185{,}604{,}309$$

with the period a little less than a quarter of the modulus.

The large size of the modulus in this equivalent algorithm provides an intuitive explanation of the high performance observed by Wichmann and Hill. More importantly, the algorithm is now in a form that is amenable to the Spectral Test. Fig. 7.1 shows the performance of the generator in comparison with two good mainframe generators chosen from Table 1 of [19], and with Knuth's $\log_2 v_k \geqslant 30/k$ criterion.

LCG(35) refers to the linear congruential generator with $a = 19935388837$, $m = 2^{35}$, and LCG(48) refers to the case $a = 31167285$, $m = 2^{48}$. The Wichmann–Hill generator (W–H) is shown here, and elsewhere in what follows, as open circles. The

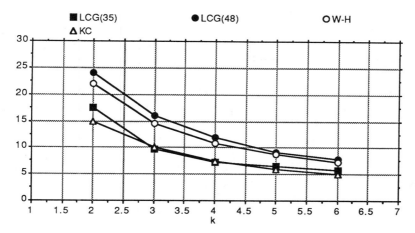

Fig. 7.1 — $\log_2 v_k$ vs k for Wichmann–Hill and mainframe generators.

lower limit of acceptable performance according to Knuth's Criterion (KC) is shown here, and elsewhere, as open triangles. Note that the mainframe generator LCG(35) barely meets Knuth's criterion, and that the Wichmann–Hill scheme (which will run happily on an 8-bit microprocessor) is far better — and almost as good as the LCG(48) mainframe generator.

7.4 OTHER WICHMANN–HILL-TYPE GENERATORS

There is another way of looking at the origin of the good performance of the Wichmann–Hill algorithm. The operation of taking the fractional part (the mod 1 operation) of a linear combination of variables distributed approximately uniformly on [0,1] induces a conspicuous improvement in the degree of uniformity. There is a kind of Central Limit process at work here, possessing remarkably rapid convergence to the uniform limiting distribution. Details of the analysis are given in [23, 24]. The outcome is that three or four generators are enough for most practical purposes. But even with only three generators, the search over the whole parameter space to find suitable moduli is a large task. We have taken a short cut by considering only primes of the form

$$p = 2p^* + 1$$
$$q = 2q^* + 1$$
$$r = 2r^* + 1$$

where p^*, q^* and r^* are again distinct primes.

For these generators, there is only a small reduction in the period, which is $2p^*q^*r = (p-1)(q-1)(r-1)/4$ compared with the maximum possible $(p-1)(q-1)$ $(r-1)$ in the case when these three terms are mutually coprime. There are also some advantages in choosing X_0.

Like Wichmann and Hill, we take the multipliers a, b and c to be the largest primitive elements (mod p, mod q and mod r, respectively) not exceeding $\sqrt{p}$, $\sqrt{q}$ and $\sqrt{r}$, respectively. There is some feeling in the random number generating community that this is a 'good thing' — but we are somewhat sceptical: perhaps other primitive elements would be suitable.

The surprising fact is that about 20% of all generators of this type, over a hundred in all (of which the original Wichmann–Hill scheme itself is an example), are 'very good' in terms of the v_k values obtained from the Spectral Test. In addition there is not much variation among the members of this set, so that the original can be regarded as a typical representative.

Should it be required, the performance can be improved by increasing the values of p, q and r (with a concomitant increase in wordlength). Further improvements are obtained by using four, rather than three, subgenerators.

In Fig. 7.2, W–H(3,8/16) refers to the original 3-component Wichmann–Hill

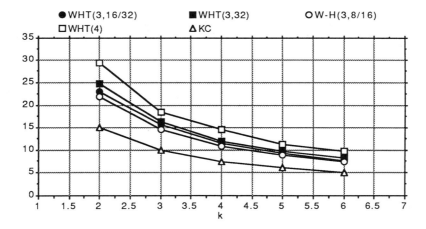

Fig. 7.2 — $\log_2 v_k$ vs k for higher-performance Wichmann–Hill types.

generator implementable in four lines of code on a 16-bit machine, or in seven lines on an 8-bit machine: the moduli are

30269,30307,30323

WHT(3,16/32) refers to a 3-component Wichmann–Hill-type generator implementable on either a 16-bit or a 32-bit machine: the moduli are

60539,60779,61343

WHT(3,32) refers to a 3-component Wichmann–Hill-type generator on a 32-bit machine in either the 'long' or the 'short' version: the moduli are

90107,91283,93179

WHT(4) refers to a typical 4-component Wichmann–Hill-type generator. As before, KC refers to the lower-limit Knuth Criterion ($30/k$).

7.5 SUMMARY AND CONCLUSIONS

We have seen that it is relatively straightforward to design generalisations of the Wichmann–Hill generator for any of the machine wordlengths that are likely to be encountered in practice. Just as important is the fact that the resulting designs **can be proved** to be good generators, by applying the Spectral Test to the equivalent single linear congruential generator.

If random sequences of **bits** are required rather than random sequences of **numbers** in the range [0,1], it is a simple matter to concatenate and serialise a suitable number of output bytes. Because the Spectral Test provides a guarantee of randomness of samples taken k-at-a-time we can be sure that provided we choose the wordlength, generator parameters and k-value appropriately, serialisation will not introduce unwanted correlation. The serial bit stream will carry the same guarantee of good performance as the generator — and we would not need to use pseudorandom multiplexing to increase the randomness.

We recommend that this technique be used in those applications requiring pseudorandom permutations for time- and/or frequency-multiplexing. Flawed generators can only weaken such systems, and many popular generators can be shown to be flawed. The generalised Wichmann–Hill generators can, on the other hand, be designed to pass the Spectral Test with flying colours to any specified level of performance, without resorting to unrealisably large wordlengths.

The situation is not quite so clear-cut for the RSA key-generation application. We have not been able to show, for instance, that a poor generator selectively skips primes, or possesses other undesirable biases that might serve as a 'signature' to assist the cryptanalyst. Even so, it would seem sensible not to risk it: the extra cost of using a generalised Wichmann–Hill generator would bestow a disproportionate increase in confidence, as these appear to be the best generators that can presently be designed.

Finally we point out that we can view the equivalence between the ultra-high performance long-period single linear congruential generator and the Wichmann–Hill composite of several poor performance short-period generators as an example of the Complexity–theoretic 'divide-and-conquer' approach to algorithm design currently popular in Computing Science [26]. We speculate that the bit-stream obtained by serialising the output of a generalised Wichmann–Hill generator has a linear complexity (defined as the length of the shortest linear feedback shift-register required to generate the bit-stream [1,14]) much in excess of the sum of the linear complexities of the (serialised) component subgenerators.

ACKNOWLEDGEMENTS

Part of this work was supported by the SERC under Contract GR/D 27211: 'Investigation and Design of Under-Specified Portable Random Number Generators with Nonuniform Power Spectra'.

REFERENCES

[1] H. Beker & F. Piper, *Cipher Systems*, Chapter 5 'Linear shift registers'; Chapter 6 'Non-linear algorithms', pp. 175–246, (1982) Northwood Publications.

[2] S. M. Edwardson, 'Scrambling and encryption for direct broadcasting by satellite', *First International Conference on Secure Communication Systems*, London, 22–24 February (1984) IEE Conference Publication No. 231, pp. 71–78.

[3] A. J. Bromfield, 'Security for voice communication'. See Chapter 8, this volume.

[4] L. C. Litwin, 'Speech scrambling — single chip LSI solutions for use in cellular and land mobile radio', *Second International Conference on Secure Communication Systems*, London, 27–28 October (1986) IEE Conference Publication No. 269, pp. 88–91.

[5] E. W. Beddoes, 'A communication security system for cellular radio', *Ibid.*, pp. 98–103.

[6] C. T. Spracklen & C. Smythe, 'Direct sequence spread spectrum access to local area networks', *Ibid.*, pp. 28–32.

[7] D. H. Crossfield & D. J. Parrish, 'Efficient microcomputer based exponentiation techniques for the RSA algorithm', *Ibid.*, pp. 58–61.

[8] C. B. Brookson & S. C. Serpell, 'Security on the British Telecom SAT-STREAM service', *First International Conference on Secure Communication Systems*, London, 22–24 February (1984) IEE Conference Publication No. 231, pp. 8–13.

[9] R. L. Rivest, A. Shamir & L. Adleman, 'A method for obtaining digital signatures and public key cryptosystems', *Comms. ACM*, **21,** No. 2, 120–126 (1978).

[10] J. Gordon, 'Implementation of the RSA', *International Conference on System Security: the Technical Challenge*, Conference Proceedings, pp. 185–195, October (1985) Online Publications.

[11] D. J. Bond, 'Practical primality testing', *First International Conference on Secure Communication Systems*, London, 22–24 February (1984) IEE Conference Publication No. 231, pp. 50–53.

[12] H. W. Lenstra Jr, 'Primality testing algorithms', *Seminaire Bourbaki*, 33e annee, (1980–1981) No. 576, pp. 576–(01/15).

[13] M. G. Kendall & B. Babbington–Smith, 'Randomness and random sampling numbers', *J. Roy. Statist. Soc. (A)*, **101,** 147–166 (1938).

[14] F. Piper, 'The use of sequences in encryption', *International Conference on System Security: the Technical Challenge*, Conference Proceedings, pp. 153–159, October (1985) Online Publications.

[15] D. E. Knuth, *The Art of Computer Programming*, Second Edition, Volume 2: *Seminumerical Algorithms*, pp. 170–173, (1983) Addison-Wesley, Reading, MA.

[16] *Ibid.*, pp. 89–113.

[17] R. Coveyou & R. MacPherson, 'Fourier analysis of uniform random number generators', *J. Assc. Comp. Mac*, **1**, 100–119 (1967).

[18] G. Marsaglia, 'Random numbers fall mainly on the planes', *Proc. Nat. Acad. Sci. (US)*, **61**, 25–28, September (1968).

[19] D. E. Knuth, *The Art of Computer Programming*, Second Edition, Volume 2: *Seminumerical Algorithms*, pp. 102–103, (1983) Addison-Wesley, Reading MA.

[20] B. A. Wichmann & I. D. Hill, 'An efficient and portable pseudo-random number generator', Algorithm AS 183, Applied Statistics, pp. 188–190, (1982).

[21] B. A. Wichmann & I. D. Hill, 'A pseudo-random number generator', National Physical Laboratory Report DITC 6/82 (1982).

[22] B. A. Wichmann & I. D. Hill, 'Building a random-number generator', *BYTE*, **12**, No. 3, 127–128, March (1987).

[23] R. C. Coates, K. V. Lever & G. J. Janacek, 'Monte Carlo simulation of communication systems', Submitted to *IEEE Proc.*, Speical Issue on Monte Carlo Simulation, to be published.

[24] G. J. Janacek & K. V. Lever, 'High-performance easily-portable uniform random number generators: a "divide-and-conquer" approach to complexity', *Fifth International Symposium on Data Analysis and Informatics*, Institut National de Recherche en Informatique et en Automatigue (INRIA), 29 September (1987) Versailles, France.

[25] T. M. Apostol, *Introduction to Analytic Number Theory*, p. 117, (1976) Springer, New York.

[26] D. F. Stanat & D. F. McAllister, *Discrete Mathematics in Computer Science*, pp. 248–256, (1977) Prentice-Hall, Englewood Cliffs, NJ.

8

Security for voice communications

A. J. Bromfield
Ernst & Whinney, Becket House, 1 Lambeth Palace Road, London, SE1 7EU,
UK

8.1 INTRODUCTION

Speech is still a very important communication method, even today with the availability of fast and effective data transmission techniques. It is particularly appropriate when a rapid exchange of information and ideas is called for. Enormous quantities of information are disseminated in spoken form over the world's communications networks.

There has long been an awareness of the security hazards of voice communications by radio, but only recently has it been generally recognised that line communications are also seriously at risk. There have been a number of recent cases in the UK of wire-tapping being used to gain a commercial or technical advantage. The advent of cellular radio and microwave radio transmission of telephone calls means that telephone conversations are no longer communicated exclusively by line. The risk of interception is therefore probably even greater in the communications networks of today than previously.

8.2 SPEECH TRANSMISSION

Speech is transmitted in a communications network in one of two fundamentally different ways. An analogue transmitter sends a continuously variable signal with limited bandwidth and amplitude. The transmitted waveform is the direct representation of the energy in the speech. A digital transmitter sends a signal which can take on only a number of discrete values. A digital signal normally consists of a waveform which represents a stream of binary digits.

In order that speech may be transmitted in digital form, it must first be coded into a digital sequence. There are many different techniques for performing such an analogue-to-digital (A/D) conversion. One such technique is known as Pulse Code

Modulation (PCM). A digital approximation to the energy in the speech signal is performed at (frequent) periodic intervals. Since an approximation must be used, the procedure is subject to what is called quantisation error. This means that when the speech is converted back into analogue form it will differ slightly from the original analogue signal. The better the approximation used, the better the quality of the speech reproduced.

The bandwidth required for the transmission of PCM coded speech at a data rate of 64 kbit/s as recommended in the CCITT standard far exceeds that of a normal speech channel. Typically, a bandwidth of 3 kHz is used for speech, but 64 kbit/s PCM requires a bandwidth of 32 kHz.

Other A/D techniques can be used to reduce the bandwidth requirement of digital speech. This can, however, only be done at the expense of the complexity of the A/D conversion algorithm. Generally speaking, there is a trade-off between bit rate, speech quality, complexity and time delay. At the present time, A/D techniques which produce a digital sequence which can be transmitted over a narrow band channel are expensive and produce speech of a synthetic quality. A vocoder is one such A/D method; it is considered in section 8.5.

This lack of a good quality low-bit-rate speech coder means that analogue scrambling techniques are still very important. The type of communications channel still has a limiting effect on the choice of a speech security system. The different generic types of speech security euqipment are defined in the next section.

8.3 SYSTEM SECURITY

There are two basic types of speech security system. An analogue scrambler transmits an analogue signal whereas a digital scrambler or speech encryptor transmits a digital signal. There are two distinct kinds of analogue scrambler. A pure analogue scrambler retains an analogue signal throughout the scrambling process. An analogue scrambler which incorporates digital signal processing performs an A/D conversion upon the speech and scrambles it in digital form, before converting it back into an analogue signal for transmission. The term 'digital scrambling' is occasionally, and rather inconsistently, applied to the latter technique. Figs. 8.1–8.3 illustrate the three different systems.

The analogue scrambler with no digital signal processing illustrated in Fig. 8.1 is the simplest and often the cheapest system available. Such a scrambler would usually be suitable for low security or privacy protection on a wide variety of communications networks.

The analogue scrambler with digital signal processing illustrated in Fig. 8.2 can be a very sophisticated system, offering a good security level. Most available scramblers fall into this category; a wide range of actual protection is offered. Such a scrambler would be suitable for most communications networks. However, its introduction into a network would almost certainly result in some loss of conveneience. One particular source of inconvenience is the time delay introduced by many scramblers, which can be comparable with that introduced by the use of a satellite link.

The digital speech encryptor illustrated in Fig. 8.3 is potentially the best method of securing speech communications. Unfortunately, the technique in its most basic form is unsuitable for many networks. However, sophisticated digital speech

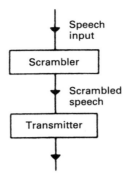

Fig. 8.1 — Analogue speech scrambler with no digital signal processing.

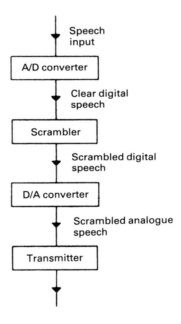

Fig. 8.2 — Analogue speech scrambler with digital signal processing.

encryptors using speech coding techniques, such as the vocoder and fast modems, are now becoming available. These systems can provide digital security over the PSTN, for example.

One of the most important properties of a speech security device is that it should transform the input speech signal into an unintelligible output signal. Unfortunately,

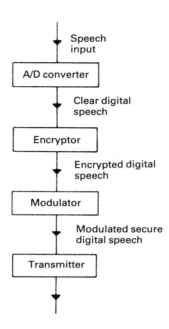

Fig. 8.3 — Digital speech encryptor.

it is extremely difficult to remove all the intelligibility from a speech signal, transmit it and yet still be able to reconstruct it in an acceptable form at the receiver. For example, the rhythm of the speech might be retained in the scrambled sound. For this reason we introduce the concept of residual intelligibility. The residual intelligibility of a scrambler is a measure of the information still present in a scrambled output from the device. Residual intelligibility is normally assessed empirically by noting the ability of a number of listeners to obtain intelligence from scrambled speech. A quantitative expression is often quoted for residual intelligibility, but this does not mean that a direct comparison between scramblers is possible. Such a comparison should only be made if the same organisation were responsible for assessing both equipments.

Note that it is a necessary but normally insufficient requirement that a scrambler has a low residual intelligibility. The true security of a sophisticated system is dependent upon a number of additional factors. There are four main factors which limit the security achievable with an analogue scrambler. Firstly, a speech signal is highly redundant in the sense that it changes very slowly and there are many pauses. An attacker can make use of this when attempting to descramble an intercepted message. Secondly, most scrambling operations have a tendency to expand the bandwidth of the analogue signal; this can mean that some of the signal is lost during transmission, and the recovered speech following descrambling is of poor quality. The third limiting factor is the noise and distortion introduced during transmission, which is often amplified by a descrambling operation. The net result of this effect is that some scramblers will work very well over a perfect channel but completely fail in poor channel conditions. The final limiting factor is that users will only tolerate a certain time delay over the channel. Security could be increased by delaying parts of the speech by several seconds, but this would introduce an unacceptable channel delay.

The redundancy of speech limits the security achievable with a digital speech encryptor. Other limitations are common to the data security problem.

A more general consideration of cryptographic security offered by either an analogue or a digital speech security system is beyond the scope of this chapter. Such considerations as the size of the key space and the algorithm used to select scrambling transformations or digital enciphering sequences are of course crucial in determining the overall security offered by a voice security system. These considerations are of course common to the data security problem.

The key management facilities offered should be adequate for the type of network for which the equipment is intended. The larger the network, the more important key management facilities become. A full discussion of key management techniques is beyond the scope of this chapter. The effectiveness of the scrambler itself should not be sacrificed for the sake of convenient key management.

Before considering which kind of speech security equipment may be suitable for a given network, it is essential that a threat assessment is performed. The organisations which provide the threat need to be identified. The level of facilities these organisations possess for interception and analysis of intercepts needs to be evaluated. Also the potential consequences of compromise must be considered.

8.4   ANALOGUE SPEECH SCRAMBLING SYSTEMS

Analogue speech scramblers transform the analogue speech (often by manipulation of a digital form of the signal). A speech signal may be considered as a three-dimensioned quantity, with amplitude, frequency and time defining the dimensions. The scrambling process may therefore be considered to be a transformation of a surface in a three-dimensional space.

For convenience, analogue scramblers are divided up once more into four basic categories: frequency scramblers, time domain scramblers, multi-dimensional scramblers and sample-based scramblers. The types of scrambling transformation employed by the different systems are briefly introduced below. Note that the classification is somewhat artificial and ignores amplitude scrambling as an explicit method. The distinction between the four types is somewhat blurred, and all scramblers may be considered to incorporate some scrambling in the amplitude dimension.

8.4.1   Frequency scramblers

This describes a class of scramblers which operates principally in the frequency domain. The scrambling operation alters the frequency components of the speech. The speech surface in three-dimensional space is dissected by planes of constant frequency (parallel to the plane defined by the time and amplitude axes) and the dissected parts rearranged. This rearrangement may be a combination of reflections and translations. The transformed surface should occupy the same bandwidth as the original surface (in practice this is never quite achieved). Three standard transformation types are introduced in Fig. 8.4, shown in two dimensions (averaged over time).

Frequency scramblers are normally implemented without converting the speech into digital form, using analogue filters. The number of available transformations in such systems is rather limited and therefore the security offered rather restricted, even if the transformation is varied with time. The advantage of such systems is that they can yield a low level of residual intelligibility, give good speech quality and introduce negligible system time delay. These properties, together with the low cost of such scramblers, make them suitable for use as privacy devices capable of deterring the casual eavesdropper in, for example, police radio networks.

Frequency scramblers implemented using digital signal processing come into the category of sample-based scramblers. A much finer dissection can be achieved with a much larger set of available transformations.

8.4.2   Time element scramblers
This describes a class of scramblers which operates principally in the time domain. The scrambling operation divides the speech into time slots called segments, and transmits them in a scrambled order. The speech surface in a three-dimensional space is dissected by constant time planes (parallel to the plane defined by the frequency and amplitude axes) and the dissected parts rearranged. The rearrangement is normally a straightforward translation, but reflections are sometimes also incorporated. The portions of speech are known as segments. The segments are usually, but not always, of constant duration. There are a number of different rearrangement strategies in common use; the hopping-window system is described below as an example.

A hopping-window system operates on groups of n consecutive segments called frames. Each segment is considered to be part of one frame only so that frames do not overlap. The speech is scrambled by permuting the segments within a frame using a permutation from the set of $n!$ available. The system is illustrated in Fig. 8.5.

The choice of segment duration T and frame length n affects the system delay $2nT$. The design of a time element scrambler involves a careful consideration of segment length, frame size, resulting system delay time, permutation choice (many permutations are ineffective) and synchronization. A typical system would use a frame length of eight segments with duration 50 ms and a system time delay of 800 ms.

Time domain scrambling techniques are potentially more effective than pure frequency domain techniques but suffer from the disadvantage that they must introduce a significant time delay into the transmission path if a low level of residual intelligibility and a high level of security are to be achieved. Their use usually results in a noticeable reduction in audio quality.

8.4.3   Multi-dimensional scramblers
This describes a class of scramblers which operates in more than one dimension, normally in both the time and the frequency domains. They normally employ a time element scrambler with some fairly simple frequency scrambling in addition, or, alternatively, a frequency scrambler with some fairly simple time element scrambling. Since time element scrambling is, generally speaking, potentially the stronger

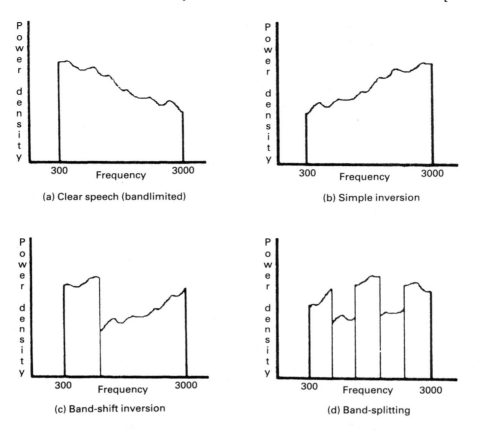

(a) Clear speech (bandlimited)

(b) Simple inversion

(c) Band-shift inversion

(d) Band-splitting

Fig. 8.4 — Some frequency scrambling techniques.

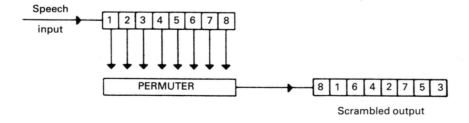

Fig. 8.5 — Hopping-window time element scrambler.

of the two operations, the former kind is normally to be preferred. A well designed multi-dimensional scrambler yields a very low level of residual intelligibility and is generally resistant to all but the most sophisticated attacks. An attack on such a device could only be successful following a considerable investment in both time and money. Unfortunately, not all scramblers which claim to be multi-dimensional and highly secure justify such a claim. Poorly designed devices, which for example use a linear sequence generator for selection of the scrambling algorithm, can lull the user into a false sense of security. Drawbacks to the multi-dimensional system are similar to those of the time element scrambler.

8.4.4 Sample-based scramblers

This describes a relatively new type of scrambling technique normally implemented using digital signal processing. Dissection of the speech surface into much smaller component pieces prior to transformation is now possible.

The problems associated with sample-based techniques are synchronisation, bandwidth expansion and lack of robustness in the presence of channel distortion. A system which is technically feasible and works very well theoretically may collapse totally in the presence of typical channel distortion. A successful sample-based technique probably requires good channel equalisation (compensation for distortion) and accurate synchronisation.

A number of systems have been proposed in the literature; see, for example, [31] and [37], [38]. Practical implementations of these systems are still quite rare, but a great deal of research is being dedicated to overcoming the problems associated with these systems, since potentially they offer greater security than the more traditional analogue scrambling techniques.

The choice of an analogue scrambler must involve a trade-off between security, price and overall performance. The security should be considered separately, as it is of prime importance.

8.5 DIGITAL SPEECH ENCRYPTORS

A digital speech encryptor actually encrypts a digitally coded speech signal in the same way that any other digitally coded data are encrypted. It does not suffer from the restriction that the encrypted signal must be converted back into analogue form prior to transmission as with an analogue scrambler incorporating digital signal processing.

Securing a digital speech network is scarcely distinct from securing any other digital network, though error propagation and block ciphers should generally be avoided. We briefly consider the problems associated with adding a digital speech encryption facility to an existing analogue network.

The A/D conversion is a relatively simple problem for a wide-band channel such as an 8 kHz VHF radio channel. On the other hand, the A/D conversion is a non-trivial problem if a narrow-band channel is to be protected. Such a channel can support an absolute maximum data rate of 9.6 kbit/s, even if an expensive modem is used. Digital speech at 2.4 kbit/s can be produced by the use of a special A/D converter called a vocoder. A vocoder computes and transmits parameters which can be used to synthesise speech at the digital-to-analogue (D/A) converter. The resulting speech sounds slightly unnatural, but word and speaker recognition can be good.

A digital speech encryptor/decryptor will therefore consist of A/D and D/A converters, a data encryptor and a modem. The performance of such a system depends upon its three components. The audio quality in good conditions will be determined by the A/D and D/A converters, whereas in poor channel conditions, all three components become significant. The security level offered is mainly dependent upon the encryptor, but the redundancy in the digital output of the A/D converter may also be significant. The residual intelligibility of the transmitted signal should be zero.

For completeness, we briefly review the properties required of the data encryptor. It should implement a secure, non-error-propagating algorithm and should have a reliable synchronisation system. The key management system should ensure that a different working key is used for every conversation. This is particularly important for speech applications because of the redundant nature of the digital speech.

8.6 CONCLUSION

There is an increasing awareness of the need to protect sensitive voice traffic in communications networks. The price and performance of speech coders for narrow-band channels mean that until these channels are replaced in digital networks world-wide, analogue scrambling techniques will continue to be important. Techniques for the security of digital speech are essentially the same as those for protecting any other kind of data.

A great variety of analogue scrambling techniques are in use world-wide. Recently, some new techniques have been developed which show a great deal of promise. This chapter has introduced some of the main techniques. The books and papers listed in the bibliography are merely a selection from the many available. The book [7] contains a much fuller bibliography.

BIBLIOGRAPHY

[1] Aegean Park Press, *Speech and Facsimile Scrambling and Decoding*, (1981) Aegean Park Press.
[2] E. W. Beddoes, 'A communication security system for cellular radio', *Proc. IEE Conf. Communications Security*, 98–103 (1986).
[3] H. J. Becker & F. C. Piper, 'Analogue speech scrambling', *New Elect.*, **15** (17), 28–32 (1982).
[4] H. J. Becker & F. C. Piper, 'Digital speech scrambling', *New Elect.*, **15** (18), 94–100 (1982).
[5] H. J. Becker & F. C. Piper, 'Communications security: a survey of cryptography', *Proc. IEE* 129 *Part A*, 357–376 (1982).
[6] H. J. Becker & F. C. Piper, *Cipher Systems: the Protection of Communications*, (1982) Van Nostrand-Rheinhold, New York.
[7] H. J. Becker & F. C. Piper, *Secure Speech Communications*, (1985) Academic Press, New York.

[8] R. M. Cosentino & S. J. Meehan, 'An efficient technique for sample-masked voice transmission', *IEEE J. Selected Areas in Comm.*, **SAC-2** (3), 426–433 (1984).

[9] R. V. Cox & J. M. Tribolet, 'Analog voice privacy systems using TFSP scrambling: full duplex and half duplex', *Bell Syst. Tech. J.*, **62** (1), 47–61 (1983).

[10] J. C. Delgado & J. M. Tribolet, 'Analog full-duplex speech scrambling systems', *IEEE J. Selected Areas in Comm.*, **SAC-2** (3), 456–459 (1984).

[11] J. L. Flanagan & M. R. Schroeder, B. S. Atal, R. E. Crochiere, N. S. Jayant and J. M. Tribolet, 'Speech coding', *IEEE Trans. Comm.*, **COM-27** (4), 710–737 (1979).

[12] A. Gersho, 'Perfect secrecy encryption of analog signals', *IEEE J. Selected Areas in Comm.*, **SAC-2** (3), 460–466 (1984).

[13] A. Gersho & R. Steele, 'Encryption of analog signals—a perspective', *IEEE J. Selected Areas in Comm.*, **SAC-2** (3), 423–425 (1984).

[14] N. S. Jayant, 'Analog scramblers for speech privacy', *Comput. Security*, **1**, 257–289 (1982).

[15] N. S. Jayant. 'Coding speech at low bit rates', *IEEE Spectrum (August)*, 58–63 (1986).

[16] N. S. Jayant, R. F. Cox, B. J. McDermott & A. M. Quinn, 'Analog scramblers for speech based on sequential permutations in time and frequency', *Bell Syst. Tech. J.*, **62** (1), 25–45 (1983).

[17] N. S. Jayant, B. J. McDermott, S. W. Christiansen & A. M. Quinn, 'A comparison of four methods for analog speech privacy', *IEEE Trans. Comm.*, **COM-29** (1), 18–23 (1981).

[18] D. Kahn, *The Codebreakers*, (1967) Macmillan, London.

[19] S. C. Kak, 'An overview of analog signal encryption', *IEE Proc.*, **130**, Part F, 399–404 (1983).

[20] S. C. Kak & N. S. Jayant, 'On speech encryption using waveform scrambling', *Bell Syst. Tech. J.*, **56** (5), 781–808 (1977).

[21] B. S. Kaliski, 'Wyner's analog encryption scheme: results of a simulation', in *Advances in Cryptology*, *Lecture Notes in Computer Science*, Vol. 196, pp. 83–94, (1985) Springer-Verlag, New York.

[22] L. G. Litwin, 'Speech scrambling—single chip LSI solutions for use in cellular and land mobile radio', *Proc. IEE Conf. Communications Security*, pp. 88–91 (1986).

[23] L. Lee & G. Chou, 'A new time domain speech scrambling system which does not require frame synchronisation', *IEEE J. Selected Areas in Comm.*, **SAC-2** (3), 443–455 (1984).

[24] N. R. F. Mackinnon, 'The development of speech encipherment', *Radio Electron. Eng.*, **50** (4), 147–155 (1980).

[25] A. M. MaCalmont, 'Measuring security in analog speech communications security devices', *Proc. IEEE Int. Conf. Commun.*, Seattle, pp. 16.5.1–16.5.4 (1980).

[26] R. E. Nelson, 'A guide to voice scramblers for law enforcement agencies', NBS Special Publication 480–8 (1976).

[27] M. V. Orceyre & R. M. Keller, 'An approach to secure voice communication

based on the data encryption standard', *IEEE Commun. Soc. Mag.*, **16** (6), 41–52 (1978).

[28] V. J. Phillips, M. M. Lee & J. E. Thomas, 'Speech scrambling by the reordering of amplitude samples', *Radio Electron. Eng.*, **41** (3), 99–112 (1971).

[29] V. J. Phillips & J. R. Watkins, 'Speech scrambling by the matrixing of amplitude samples', *Radio Electron. Eng.*, **43** (8), 459–470 (1973).

[30] F. Pichler 'Analog scrambling by the general fast Fourier transform', in *Cryptography, Lecture Notes in Computer Science*, Vol. 149, pp. 173–178, (1982) Springer-Verlag.

[31] K. Sakurai, K. Koga & M. Murantani, 'A speech scrambler using the fast Fourier transform technique', *IEEE J. Selected Areas in Comm.*, **SAC-2** (3), 434–442 (1984).

[32] M. R. Sambur & N. S. Jayant, 'Speech encryption by manipulations of LPC parameters', *Bell Syst. Tech. J.*, **55** (9), 1373–1388 (1976).

[33] N. J. A. Sloane, 'Encrypting by random rotations', in *Cryptography, Lecture Notes in Computer Science*, Vol. 149, pp. 71–128, (1982) Springer-Verlag, New York.

[34] K. P. Timmann, 'The rating of understanding in secure voice communication systems', in *Cryptography, Lecture Notes in Computer Science*, Vol. 149, pp. 157–164, (1982) Springer-Verlag, New York.

[35] S. Udalov, 'Microprocessor-based techniques for analog voice privacy', *Proc. IEEE Int. Conf. Commun.*, Seattle, pp. 16.2.1–16.2.6 (1980).

[36] P. K. S. Wah & P. Schobi, 'A study of Wyner's analog speech scrambling schemes', *Proc. Int. Carnahan Conf. Security Technology*, Zurich, pp. 109–116 (1983).

[37] A. D. Wyner, 'An analog scrambling scheme which does not expand bandwidth, Part I: Discrete time', *IEEE Trans. Inform. Theory*, **IT-25** (3), 261–274 (1979).

[38] A. D. Wyner, 'An analog scrambling scheme which does not expand bandwidth, Part II: Continuous time', *IEEE Trans. Inform. Theory*, **IT-25** (4), 415–425 (1979).

9

A communication channel model of cryptography: a unified view of codes and ciphers

G. Davies
Royal Navy, *HMS Thunderer*, Royal Naval Engineering College, Plymouth, UK

9.1 INTRODUCTION

A message in a language may be considered as a finite sequence of characters selected from a finite alphabet according to certain rules. This definition embraces a spectrum ranging from real languages, such as English, where the rules are complex and intuitive and can be represented only statistically, to mathematical languages such as error correcting codes, where the rules are precise and deterministic and can be implemented by simple logic circuits.

All such languages share the common feature that the selection constraints impose a structure on the messages: their character sequences are recognisably non-random. This intrinsic pattern gives them a degree of resistance to corruption. The corruption may be from a noisy communication channel or an enciphering key: in either case, pattern recognition techniques may be used in an attempt to recover the original message.

Thus structure in the source language may be exploited by the communication engineer to increase the reliability of messages, but also by the cryptanalyst to attack a cipher. Reliability and security appear, therefore, as conflicting goals with a potential trade-off between them. We shall, however, deduce an important and surprising result: it is possible, in principle, to achieve total reliability and total security, sumultaneously, over a noisy communication channel.

Structure also entails redundancy, since the selection constraints confine the message sequences of a given length to a small subset of the possible unconstrained sequences of that length. Thus structure, and the consequent redundancy, reduce the rate of the language — the average information per character it can carry — and imply a further potential trade-off.

These inter-relationships between reliability, rate, security and redundancy unite the fields of coding and cryptography. In order to illustrate, and hopefully to exploit, this unity we first recall some basic definitions from Information Theory.

9.2 INFORMATION THEORY

Consider message sequences X, of length N, selected from an alphabet of L characters. The entropy $H(X)$ of the messages — the average uncertainty as to which messages were selected — is given by

$$H(X) = E[-\log_2 p(X)] \text{ bits} \tag{1}$$

where $p(X)$ is the *a priori* probability of selecting the message X, E denotes the expectation value, and the averaging process is taken over all possible messages of length N.

The rate R of a language is defined as the entropy of its messages, normalised to a per-character basis:

$$R = \frac{H(X)}{N} \text{ bits per character} \tag{2}$$

where the sequence length N is made long enough to ensure that the messages embrace all the structural constaints of the language.

If, instead of messages, we consider random sequences of a given length — the N characters being selected equiprobably and independently — then the entropy and rate assume the absolute values H_0 and R_0, respectively, given by

$$H_0 = \log_2 S \text{ bits}, \quad R_0 = \log_2 L \text{ bits per character} \tag{3}$$

where S is the total number of unconstrained sequences of length N. Clearly, $S = L^N$.

The redundancy D of the language is now defined by

$$D = R_0 - R \text{ bits per character} \tag{4}$$

The redundancy D is therefore a simple quantifier of the degree of structure in a language, since it measures the amount by which the rate of the meaningful message sequences falls below that of the unconstrained random sequences, and structure, or pattern, is simply deviation from randomness.

In geometrical terms it also quantifies the average separation of message sequences in the space of all possible sequences — the Hamming space of the language. Thus if we tessellate this space with cells that each contain only a single message, and use minimum distance decoding, then the greater this separation, the greater the corruption the system can withstand. This again illustrates the fact that high reliability goes hand-in-hand with high redundancy and low rate.

We now consider our messages to be transmitted over a noisy communication channel or enciphered with a secret key — the same diagram (Fig. 9.1) serves for both — to give the corrupted sequence Y, also of length N.

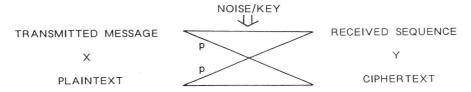

Fig. 9.1 — Description of message transmission.

The equivocation $H(X|Y)$ of the channel — the average uncertainty as to which message X was transmitted or enciphered, knowing which sequence Y was received or intercepted — is given by

$$H(X|Y) = E[-\log_2 p(X|Y)] \text{ bits} \tag{5}$$

where $p(X|Y)$ is the *a posteriori* probability of the message X knowing the sequence Y.

On the basis that we receive information only in circumstances in which our uncertainty is reduced, we now define the mutual information $I(X,Y)$ of the system as the difference between the *a priori* and *a posteriori* uncertainties:

$$I(X,Y) = H(X) - H(X|Y) \text{ bits} \tag{6}$$

The equivocation quantifies the corrupting power of the noise, or key, and exactly equals the information lost by its randomising action. This information is recoverable, of course, if the valid receiver has an exact copy of the noise, or the enciphering key — this latter being the usual case in a cryptographic system.

In a communication system, no such exact copy of the noise exists. However, if the equivocation of the channel can be reduced to zero, no uncertainty would remain at the receiver as to which message X was selected at the transmitter. The information obtained would then equal the entropy of the source message, since all the *a priori* uncertainty would be resolved on the receipt of Y. This is the ultimate goal of reliable communication.

Alternatively, viewed as a cryptographic system, if the equivocation can be made as large as the source entropy, then no information can be obtained from interception of the ciphertext. The *a posteriori* uncertainty as to which plaintext message X was selected — knowing the ciphertext Y — equals the *a priori* uncertainty. No uncertainty is resolved and therefore no information gained. This is the ultimate goal of secure communication.

One final parameter from basic Information Theory is the channel capacity C, defined as the maximum value of the mutual information, for a given channel, and normalised — like the rate R — to a per-character basis. Thus:

$$C = \frac{\max [I(X|Y)]}{N} \text{ bits} \tag{7}$$

where the maximum is taken over all possible source languages that could use the channel.

All of these parameters assume particularly simple values for the binary symmetric channel.

9.3   BINARY SYMMETRIC CHANNEL

In a binary symmetric channel, the noise is represented by a symmetrical corruption probability p, and the source language has only two characters ($L = 2$).

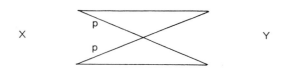

To create a language structure in the source, we consider the messages X to be error correcting codes N characters in length, of which M are unconstrained — they occur equiprobably and independently — and the remainder are totally determined by the coding algorithm, and therefore have zero uncertainty.

UNSTRUCTURED SOURCE   STRUCTURED MESSAGE

Applying our previous definitions for the entropy, rate and redundancy of a structured language to these binary error correcting codes, we have

$$H(X) = M \text{ bits}, \quad R = \frac{M}{N} \text{ bits per character} \tag{8}$$

$$R_0 = \log_2 L = 1 \text{ bit per character, since } L = 2 \text{ in this case} \tag{9}$$

$$D = R_0 - R = \frac{N - M}{N} \text{ bits per character} \tag{10}$$

and for the channel itself:

$$H(X|Y) = N[-p\log_2 p - (1-p)\log_2(1-p)] \text{ bits} \tag{11}$$

$$C = 1 - [-p\log_2 p - (1-p)\log_2(1-p)] = 1 - H(p) \text{ bits per character} \tag{12}$$

where $H(p)$ is the binary entropy function defined by

$$H(p) = -[p\log_2 p + (1-p)\log_2(1-p)], \quad H(0) = H(1) = 0 \tag{13}$$

We are now ready to deduce the most fundamental result in communication engineering: Shannon's Coding Theorem.

9.4 SHANNON'S CODING THEOREM

Consider an unconstrained stream of binary characters transmitted over a binary symmetric channel, having corruption probability p and capacity C, with N characters passed in unit time. Clearly, NC bits of information are transmitted in this time. We now impose a language structure on this stream by breaking it into M character blocks and embedding them in N character codewords: the redundant characters being totally determined by the coding algorithm. A decoding algorithm reverses this process at the receiver to recover the original unconstrained data stream.

We now consider an extended channel, with corruption probability q and capacity C', which incorporates this coding and decoding process:

C'

M characters   o————————————————————————————o   M characters

N characters   o————————————o   N characters

C

Since MC' bits of information are transmitted over the extended channel in the same unit time that NC bits of information are transmitted over the original channel, we have

$$MC' = NC \tag{14}$$

which becomes, using equation (8):

$$R = \frac{C}{C'} \tag{15}$$

Now C and C' are the channel capacities of two binary symmetric channels with corruption probabilities p and q, respectively. Therefore, using equation (12):

$$R = \frac{1 - H(p)}{1 - H(q)} \tag{16}$$

Thus, imposing a language structure — defined by the rate R — on the source data transforms the primary channel, with corruption probability p, into a modified channel with corruption probability q, these three parameters being related by equation (16).

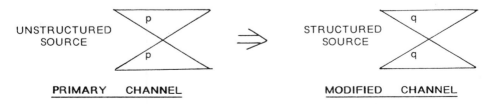

UNSTRUCTURED   p   STRUCTURED   q
  SOURCE   ⟹   SOURCE
                    p   q

PRIMARY   CHANNEL   MODIFIED   CHANNEL

Equation (16) is the central result of this chapter. It defines the achievable limits for the various trade-offs that are possible between rate, redundancy, security and reliability, and may be applied in both coding theory and cryptography.

In coding theory, it is the primary channel that is given, and its specified corruption probability p that forms the starting point for applying equation (16). The unstructured source has a rate of unity and zero redundancy. Introducing an error correcting code creates a structured binary language with a non-zero redundancy and a rate less than unity. The degree of this redundancy is within the communication engineer's control.

Equation (16) then gives the optimum trade-off that is achievable between R and q for the modified channel. It defines the minimum structure necessary — and hence the maximum rate possible — to reduce the corruption probability to q.

In cryptography, it is the structure in the source plaintext language that is given, and its specified rate and redundancy that form the starting point for applying equation (16). Introducing a secret key randomises the plaintext language, with a corruption probability p, and reduces its structure, or pattern. The degree of this randomisation is within the cryptographer's control.

Equation (16) then gives the optimum trade-off that is achievable between the primary corruption probability p, of the plaintext and its residual corruption probability q after a cryptanalyst has attacked the cipher with pattern recognition techniques. It defines the minimum randomisation necessary — and hence the minimum key required — to ensure a residual corruption probability of q.

In both cases, Nq is the radius of a hypersphere in the Hamming space of the source language that contains all the possible messages that could be decoded, or decrypted, from a given corrupt sequence. If the decoding, or decryption, is unique, the sphere shrinks to zero radius centred on the true message, Hence q is zero.

Since equation (16) defines an optimum situation, any trade-off below this value is also achievable, giving the inequality

$$R \leq \frac{1 - H(p)}{1 - H(q)} \tag{17}$$

For the communication engineer the most important consequence of this result is that no matter how noisy the given channel — defined by its corruption probability p — the modified channel may be made completely noiseless: q may be reduced to zero. For such a noiseless channel, the binary entropy function $H(q)$ is also zero, and inequality (17) reduces to

$$R \leq 1 - H(p) \quad \text{or, in general} \quad R \leq C \tag{18}$$

This last result is Shannon's Coding Theorem, which states that there exist codes of rate R, not exceeding the channel capacity C, for which totally reliable communication is possible. For rates exceeding C, however, uncorrectable errors must occur, as quantified by inequality (17).

The trade-off implicit in inequality (17) may also be used to reduce structure already present in the source data. This is useful in cryptography since it reduces the amount of key necessary to corrupt a plaintext message — clearly the weaker the original pattern, the less randomness need be introduced to remove it.

A simple way to achieve this is to use an error correcting code in reverse: the decoding algorithm is used to reduce an M character block of data to an N character block. Since M is now greater than N, the rate R, from equation (8), exceeds unity, and from equation (16) the corruption probability q of the modified channel is greater than the corruption probability p of the original channel. Considering the channel as an encryption process, and the noise as a key, the corrupting power of the cipher has been enhanced — less key is required for the same degree of security.

The reverse use of an error correcting code does not necessarily introduce corruption at the valid receiver — which, after decryption, uses the coding algorithm in the forward sense to recover the message — provided sufficient structure remains in the source to eliminate the errors the extra noise introduces.

The method is of particular use when there is more structure — redundancy — in the original source language than is strictly necessary to correct all the corruption errors over the given channel. We are here optimising the trade-off between security and reliability in communication.

To pursue the communication channel model of cryptography further, we introduce the most fundamental result in cipher theory — the Unicity Distance Theorem.

9.5 UNICITY DISTANCE THEOREM

We take as our model of a cryptographic system the substitution cipher, in which the key is simply added to the plaintext to obtain the ciphertext:

$$\text{CIPHERTEXT} = \text{PLAINTEXT} + \text{KEY} \quad (\text{modulo } L) \tag{19}$$

If the key is shorter than the plaintext message length, it is simply repeated. In this case the key entropy is finite, as no further uncertainty is introduced after the first period. If the key length always equals the plaintext length, no matter how great, the cipher is known as a one-time pad — in this case the key entropy increases with the message length and has no upper limit.

In principle it is possible for a cryptanalyst to test all possible keys on an intercepted ciphertext. This process must, at some stage, recover the plaintext message. How can a cipher remain secure in the face of such a theoretical attack?

To answer this question we first calculate the probability that some random process — such as a monkey pressing keys on a typewriter — might generate a message in a language simply by chance.

For a particular plaintext language, let

S_A = number of message sequences of length N
S_B = number of possible unconstrained sequences of length N

then from the definitions of R and R_0 and equation (3) we have

$$\text{Entropy of message sequences} = RN = \log_2 S_A \; ; \quad \text{therefore } S_A = 2^{RN} \tag{20}$$

Entropy of possible sequences $= R_0 N = \log_2 S_B$; therefore $S_B = 2^{R_0 N}$

$$(21)$$

where we have assumed that sufficiently long messages are equiprobable.

Thus the probability P_C of a random process generating a message of length N, in a language with a redundancy D, is given by

$$P_C = \frac{S_A}{S_B} = 2^{-[R_0 - R]N} = 2^{-DN} \qquad (22)$$

As we might expect, this probability falls exponentially with increasing message length N — monkeys type short words occasionally, long words rarely and the works of Shakespeare only in science fiction novels. The probability also falls exponentially with increasing redundancy D — the greater the pattern in a language, the less likely it is that a chance process can reproduce it. Alternatively, if a language has zero redundancy then every character sequence qualifies as a message and P_C is unity.

Now the entropy of the key $H(K)$ and the number of possible key sequences S_K are related by

$$H(K) = \log_2 S_K; \quad \text{therefore} \quad S_K = 2^{H(K)} \qquad (23)$$

again on the worst case assumption — from the cryptanalyst's point of view — that the possible key sequences are equiprobable.

Thus even if a random sequence was originally encrypted, instead of a true message, operating on the ciphertext with all possible keys will generate, by chance alone, S_C messages, where

$$S_C = S_K P_C = 2^{H(K)} . 2^{-DN} = 2^{[H(K) - DN]} \qquad (24)$$

Now, if the entropy $H(K)$ of the key is finite, there will be a length N_0 of ciphertext for which

$$H(K) - DN_0 = 0 \qquad (25)$$

Thus, from equations (24) and (25), we see that, for values of N greater than N_0, operating on a real ciphertext with all possible keys will generate, on average, less than one message by chance alone. But since one additional message will also be revealed by this process — the one originally encrypted — the true plaintext can be identified. In this case, therefore, the cipher is insecure.

On the other hand, for values of N less than N_0, operating on the ciphertext with all possible keys will generate many messages by chance alone. The true plaintext will then be but one of an ensemble of possibilities, and not unambiguously identifiable.

Even though the true message has been generated by this process, an irreducible uncertainty remains. In this case, therefore, we have progressively greater security as N decreases below N_0, since there is then an increasing number of alternative — and possible — plaintext messages. As Father Browne ònce said — 'the safest place to hide a leaf is in a forest'.

N_0 is known as the Unicity Distance of the cipher, and from equation (25) it is equal to the entropy of the key divided by the redundancy of the plaintext language;

$$N_0 = \frac{H(K)}{D} \qquad (26)$$

The Unicity Distance Theorem states that, for a ciphertext of length N, if

$$N > N_0 \qquad (27)$$

the cipher is theoretically insecure. As N decreases below N_0, however, the cipher becomes progressively more secure.

So far we have only shown that, for a plaintext message of a given length N, there is a trade-off between the key entropy — the randomising power of the cipher — and the degree of security. In coding theory we saw that the ultimate goal of totally reliable communicatin was achievable. In cryptography is the corresponding goal of totally secure communication also achievable? To answer this question we examine one particular substitution cipher: the random one-time pad.

9.6 RANDOM ONE-TIME PAD

In this cipher, the key sequence is always of the same length N as the plaintext message — no matter how long this may be — and its characters are chosen equiprobably and independently. From equation (3), therefore, the key entropy $H(K)$ is given by

$$H(K) = R_0 N \qquad (28)$$

The total number S of messages, obtained by chance alone, when decrypting the cipher text with all possible keys, is given by substitution this value of $H(K)$ into equation (24):

$$S_C = 2^{[H(K) - DN]} = 2^{[R_0 - D]N} = 2^{RN} \qquad (29)$$

where we have also substituted for the redundancy D of the plaintext language from equation (4).

But, by equation (20), this value of S_C is exactly equal to the total number S_A of messages, of length N, in the plaintext language.

Operating on the ciphertext with all keys has simply generated every message of equivalent length in the plaintext language, and all we know is that the true message is one of these. But we knew this already: our *a posteriori* uncertainty as to the correct plaintext is equal to our *a priori* uncertainty. Since information is resolved uncertainty, the information gained by applying to the ciphertext all possible pattern recognition techniques — and unlimited computer power — is zero: the message remains secret. Thus total security of communication, as well as total reliability, is achievable in principle.

It is important, however, that the one-time pad is truly random, with zero redundancy and no structure. A running text one-time pad — in which the plaintext is enciphered with another message in the language — is not totally secure. In fact it usually lies beyond the unicity distance and thus has no degree of theoretical security at all.

There is simply too much redundancy — insufficient randomness — in the enciphering key to destroy the structure in the plaintext message, leaving it vulnerable to pattern recognition cryptanalysis.

The case of a running text one-time pad will be useful in another context, so we shall examine it in more detail.

9.7 RUNNING TEXT ONE-TIME PAD

In this cipher, the key sequence is always of the same length N as the plaintext message — without cyclic repetition — but its characters are not chosen equiprobably and independently: they also form a message in the plaintext language.

In this case, therefore, the key entropy $H(K)$ is given, from equation (2), by

$$H(K) = RN \tag{30}$$

The total number S_C of messages, obtained by chance alone, when decrypting the ciphertext with all possible keys, is given by substituting this value of $H(K)$ into equation (24):

$$S_C = 2^{[H(K)-DN]} = 2^{[R-D]N} = 2^{[2R-R_0]N} \tag{31}$$

where again we have substituted for the redundancy D of the plaintext language from equation (4).

Examining the last exponent in equation (31), we see that if the rate R, and absolute rate R_0, of the plaintext language are such that

$$2R < R_0 \tag{32}$$

then operating on the ciphertext with all possible keys will generate, on average, less than one message by chance alone. Since the true plaintext message will also be revealed by this process, it will be easily identified. A running text one-time pad cipher in a language obeying inequality (32) is thus insecure — a result we shall exploit later in an error correction system known as soft decision decoding.

The rate R, and absolute rate R_0, of English is given by

$$R_0 = \log_2 L = \log_2 26 = 4.7 \text{ bits per character}$$
$$R = 1.2 \text{ bits per character}$$

These values obey inequality (32) and thus running text one-time pad ciphers in English are theoretically insecure.

Information theory, and the various trade-offs we have discussed, reveal coding theory, whose fundamental theorem is Shannon's Coding Theorem, and cryptography, whose fundamental theorem is the Unicity Distance Theorem, to be but two faces of one coin. To reinforce this unity we will show that the theorems themselves are but two complementary aspects of one general theorem.

To this end we shall derive Shannon's Coding Theorem from the Unicity

Distance Theorem, and the Unicity Distance Theorem from Shannon's Coding Theorem.

9.8 SHANNON'S THEOREM FROM THE UNICITY DISTANCE THEOREM

In a noisy binary symmetric channel, the received sequence is the addition of the transmitted message and a binary error sequence — the corrupting noise. This error sequence has a binary 'one' at every position destined to be received corrupt and a binary 'zero' at every position destined to be received correct.

$$\text{RECEIVED SEQUENCE} = \text{TRANSMITTED MESSAGE} + \\ + \text{NOISE (modulo 2)}$$

(33)

This is clearly a special case of equation (19) and permits us to regard the communication channel as an encryption process with a one-time pad key given by the binary error sequence. The probability of a binary one at any position in this sequence is equal to the corruption probability p of the communication channel.

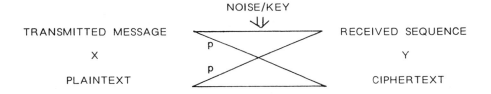

The requirement for totally reliable communication over a noisy channel is equivalent, in the encryption analogy, to a cipher being operated beyond its unicity distance. Then, by the Unicity Distance Theorem, the cipher is breakable by pattern recognition techniques and the message recoverable — hence error-free communication.

For a message of length N, the condition for operation beyond the unicity distance is

$$N \geqslant \frac{H(K)}{D}$$

(34)

The enciphering key, or corrupting noise, is also of length N and has a probability p of a binary one in any position and $(1-p)$ of a binary zero in that position — all positions being independent of each other for white noise. Its entropy $H(K)$ is thus given, from equation (1), by

$$H(K) = N[-p\log_2 p - (1-p)\log_2(1-p)] = NH(p)$$

(35)

The absolute rate R_0 of a binary language ($L = 2$) is given by

$$R_0 = \log_2 L = 1 \text{ bit per character}$$

(36)

Thus if the rate of the source — plaintext or transmitted message — is R then its redundancy D is given by

$$D = R_0 - R = 1 - R \tag{37}$$

Substituting these values for the key entropy $H(K)$ and plaintext redundance D into inequality (34), we have, after cancelling N and rearranging:

$$R \leqslant 1 - H(p) \tag{38}$$

The RHS of this inequality is the capacity C of a binary symmetric channel; thus the condition for the cipher to be breakable is:

$$R \leqslant C \tag{39}$$

But this is the Shannon Coding Theorem condition for totally reliable communication, derived as a direct consequence of the Unicity Distance Theorem.

If a cipher can be broken, and the plaintext recovered, then the uncertainty introduced by the key has been reduced to zero. Under the correspondingly similar conditions, therefore, the corrupting influence of the noise on a communication channel — the equivocation — can also be reduced to zero, and all errors in the message corrected.

Thus considering our model in turn as an encryption process and as a noisy communication channel, we see that the condition that a cipher, in the former, should be breakable is exactly equivalent to the condition that communication, in the latter, should be totally reliable. The pattern already present in a plaintext language, or that we have created in an error correcting code, is sufficient to withstand the randomising effect of the key, or of the channel noise. A cipher operated beyond its unicity distance and a code operated below its Shannon limit are one and the same thing.

This is the philosophy of the error correction technique known as soft decision decoding, which will be discussed as an illustrative example later.

Although this result was derived in the context of a binary channel, it could equally well be obtained for a multi-level channel, representing a source language with an arbitrary number of characters L. In that case, inequality (39) would be obtained directly, rather than via the special case of inequality (38). To illustrate this point, we now derive the converse result — the Unicity Distance Theorem from Shannon's Coding Theorem — for just such a multi-level channel.

9.9 UNICITY DISTANCE THEOREM FROM SHANNON'S CODING THEOREM

In a substitution cipher, the ciphertext is the addition of the plaintext and the enciphering key:

$$\text{CIPHERTEXT} = \text{PLAINTEXT} + \text{KEY} \quad (\text{modulo } L)$$

This is equivalent to a multi-level communication channel with L levels, the key being the corrupting noise:

RECEIVED SEQUENCE = TRANSMITTED MESSAGE
+ NOISE   (modulo L)

Thus such a channel forms a model of the encryption process — in both cases the character set is randomly mapped onto itself.

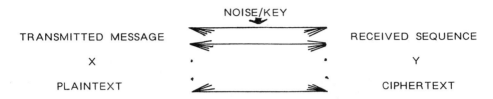

If the key is shorter than the message, the noise is periodic, with period equal to the keyword length, and has a finite entropy — no further uncertainty is added after the first cycle, of the key is a one-time pad, the noise is non-repetitive and its entropy grows with the message length.

The requirement that the cipher should have a measure of security is equivalent, in the communication analogy, to the channel being operated beyond its Shannon limit $(R > C)$. In that case, uncorrectable errors must occur at the receiver despite the redundancy of the transmitted message. The corruption probability of the modified channel — which includes the pattern recognition process — cannot be reduced to zero. These irreducible errors grow rapidly as we proceed beyond the Shannon limit — as shown, for the binary case, by equation (16) — giving increasing security for the cipher.

By Shannon's Coding Theorem, therefore, the condition for a cipher, considered as a communication channel, to have some degree of security is

$$R > C \tag{40}$$

where R is the rate of the plaintext language — the transmitted message — and C is the channel capacity.

From equations (7) and (6) this capacity C is given by

$$C = \frac{\max [I(X|Y)]}{N} = \frac{\max [H(X) - H(X|Y)]}{N} \tag{41}$$

where $H(X)$ is the source entropy and $H(X|Y)$ is the channel equivocation.

The uncertainty introduced by the corrupting sequence — noise or key — is identical in the two representations of our model: it is the same sequence. Therefore the equivocation, which quantifies the channel noise, and the key entropy, which quantifies the randomising power of the cipher, are equal:

$$H(X|Y) = H(K) \tag{42}$$

If R_0 is the absolute rate of the source language then, by equation (3), we have

$$\max H(X) = R_0 N \tag{43}$$

Thus substituting from equations (43), (42) and (41) into inequality (40) gives

$$R > R_0 - \frac{H(K)}{N} \tag{44}$$

and remembering that the difference between R and R_0 quantifies the source redundancy D, we have, after rearrangement:

$$N < \frac{H(K)}{D} \tag{45}$$

But this is the Unicity Distance Theorem condition that a cipher should be secure, derived as a direct consequence of Shannon's Coding Theorem.

If a communication channel introduces uncorrectable errors, so that messages are irreversibly corrupted, then the uncertainty introduced by the noise cannot be reduced to zero. Under the correspondingly similar conditions, therefore, the corrupting influence of a key cannot be entirely eliminated — the plaintext retains a measure of uncertainty, and hence of security.

The further we drive a channel beyond its Shannon Limit, the greater the corruption and, correspondingly, the further below its unicity distance we operate a cipher, the greater the security.

Thus considering our model in turn as a noisy communication channel and an encryption process, we see that the condition that messages, in the former, should be irreversibly corrupted is equivalent to the condition that plaintext, in the latter, should be secure. The pattern we have created in an error correcting code, or that was already present in a plaintext language, is insufficient to withstand the randomising effect of the channel noise, or of the key. A code operated beyond its Shannon limit and a cipher operated below its unicity distance are one and the same thing.

We may exploit this correspondence to produce a novel cipher system that does not require the prior exchange of secret keys between the legitimate user. This will be discussed as an illustrative example later.

The results derived so far — particularly those of this and the last section — permit us to view coding and cryptography as a coherent whole. We now formulate this as a general principle.

9.10 CODING AND CRYPTOGRAPHY FROM A UNIFIED VIEWPOINT

The problems addressed by the communication engineer and the cryptanalyst are essentially the same. Both employ pattern recognition techniques in an attempt to recover a corrupted message. For the communication engineer, the protagonist is Nature who has randomised the message with channel noise; for the cryptanalyst, the protagonist is the cryptographer who has randomised the plaintext with a secret key.

Coding and cryptography operate at opposite ends of one continuous spectrum. In our binary model this spectrum is represented by the corruption probability q of the modified channel — which incorporates the coding and decoding algorithms for error correction, and all the intrinsic structure of the plaintext language.

It is important to remember that q is the residual — and thus irreducible —

corruption probability remaining after all possible pattern recognition techniques have been applied to exploit the redundancy of the source. It is related to the primary corruption probability p of a noisy communication channel, or an enciphering key, through equation (16) — see Fig. 9.2.

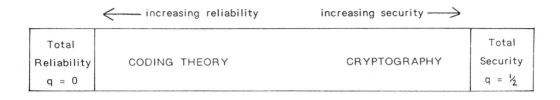

Fig. 9.2 — The corruption spectrum.

The aim of the communication engineer is to operate as close as possible to the lower end of this spectrum — zero residual corruption ($q = 0$). The aim of the cryptographer is to operate as close as possible to the upper end — total residual corruption ($q = \frac{1}{2}$). These endeavours are governed by a single theorem, which is known as Shannon's Coding Theorem when viewed from the coding end, and as the Unicity Distance Theorem when viewed from the cryptographic end.

This theorem quantifies, for a channel model with corruption probability p, the degree of pattern that must be created — or destroyed — in the source language to achieve a required level of reliability — or security — as defined by the residual corruption probability q.

This trade-off between source pattern and residual corruption probability may be given a geometrical interpretation in a Hamming space whose points represent N-sequences.

9.11 GEOMETRICAL INTERPRETATION OF PATTERN RECOGNITION

In coding theory we start with an unstructured source and a communication channel with corruption probability p. Since the source language has zero redundancy, every point in its Hamming space is a message. An N-sequence received over this noisy channel will also be a message, but not necessarily the one transmitted. A sphere of uncertainty, with radius Np, about the point representing this corrupt sequence will contain all the possible messages to which it could be decoded.

As we create increasing structure in the source language — by applying a coding algorithm — the messages move apart in the Hamming space, since an increasing number of the intervening points no longer represent permitted sequences: they do not have the correct character pattern to qualify as messages. Thus the number of possible messages within the sphere of uncertainty, to which our corrupt sequence could be decoded, falls: the ambiguity is progressively resolved and the reliability increased. We are approaching the Shannon limit from the high corruption end.

Equivalently, we may keep the message separation constant throughout this process and allow the sphere of uncertainty to shrink to a residual radius Nq, where q is given by equation (16). Again the number of possible decodings progressively falls.

When the pattern in the source language is sufficiently great, this sphere will be small enough to contain only a single message, all other points within it representing non-permitted sequences. The sphere of uncertainty then effectively collapsed to a single point, the true message — q is zero and we have total reliability. This corresponds to operation below the Shannon limit.

In cryptography we start with a structured plaintext language and a key entropy equivalent to a channel corruption probability p. Since the source language has an intrinsic pattern, the valid messages are widely scattered among invalid N-sequences in its Hamming space. An intercepted ciphertext will correspond to a point in this space. A sphere of uncertainty about this point, with radius Np, will contain all the possible plaintext messages to which this sequence could be decrypted.

Thus if the key entropy — and hence the channel equivalent p of its corrupting power — is too small, this sphere may contain only a single valid message: the cipher is insecure. This corresponds to operation beyond the unicity distance.

Before the plaintext is enciphered, however, we may remove structure from the source language, by, for example, the reverse use of an error correcting code. The distance between messages then decreases, since more of the intervening sequences now conform to the less stringent pattern constraints. Thus the number of plaintext messages within the sphere of uncertainty, to which our intercepted ciphertext could be decrypted, increases. Since ambiguity now exists as to the true message, there is a measure of security. This corresponds to operation below the unicity distance.

Equivalently, we may keep the message separation constant throughout this process and allow the sphere of uncertainty to grow to a final radius Nq, where q is given by equation (16). In applying this equation we notice that, since the source redundancy has been reduced, the rate R has increased and q is now greater than p. Again the number of possible decryptions increases.

When the pattern in the plaintext language is sufficiently small, the sphere of uncertainty will be large enough to encompass every message in the Hamming space. The ambiguity is then total. Since the *a posteriori* uncertainty as to the true plaintext is now equal to the *a priori* uncertainty, no information has been gained by a pattern recognition attack on the ciphertext — q is equal to one half and we have total security.

The analogy between coding and cryptography, besides being interesting from a theoretical and pedagogical point of view, has practical consequences. It permits cryptographic techniques to be used in error correction and coding techniques to be used in cryptography. We now give an example of each of these transposed uses.

9.12 SOFT DECISION DECODING

Our channel model permits us to view a digital communication system as an encryption process, where a binary source is enciphered with a binary key — the channel noise. The key will therefore be a sequence of ones and zeros, with the ones corresponding to those positions in the transmitted message that are destined to be received corrupt, and the zeros corresponding to those positions that are destined to be received correct. If this sequence is representative of the noise on a real channel, however, it will be very far from being totally random.

For a sequence to qualify as totally random, its uncertainty, or entropy, must be a maximum. This requires that the characters occur equiprobably and independently in the sequence — giving, in the binary case, the same distribution for ones and zeros as that for heads and tails in successive tosses of a fair coin. Any departure from either equiprobability or independence will impart structure, and hence a degree of non-randomness, to the sequence.

In a real digital channel the corruption probability p is very low; hence the ones and zeros in the noise sequence are not equiprobable. Real channels are subject to error bursts — the noise is correlated — hence the ones and zeros are not independent. These departures from total randomness are equivalent to pattern, and permit us to view the noise, in addition to the coded source, as a structured language with a definite rate and redundancy.

In cryptographic terms, therefore, this noisy channel may be considered as a running text one-time pad cipher, in which a structured plaintext source is encrypted with a structured key. For such a case we saw that, provided the intrinsic pattern was sufficient, the cipher was insecure — in the communication analogue, therefore, the possibility of error-free reception exists.

The theory of soft decision decoding — but not necessarily the practice — is to decode a received N-sequence with all possible N-sequence keys that conform to the known statistical structure of the channel noise. We then search through these decryptions to find sequences that possess the known statistical structure of the source. If this process yields a unique result — the true transmitted message — then totally reliable communication is achieved.

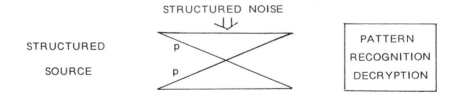

The redundancy D, and the rate R, of the structure source are related, in the binary case, by

$$D = 1 - R \qquad (46)$$

The degree of this redundancy is within the communication engineer's control by the use of an appropriate source coding algorithm.

Similarly the redundance D', and the rate R', of the structured noise are related, again for the binary case, by

$$D' = 1 - R' \qquad (47)$$

The degree of this redundancy is determined by the channel.

We now consider the binary noise as an N-sequence one-time pad key, where N is the length of the transmitted message. Its entropy $H(K)$ is given, from the definition of rate, by

$$H(K) = R'N \tag{48}$$

Let S_C be total number of chance messages — sequences that pass, by chance alone, every statistical test of being source codewords – obtained by decrypting the received N-sequence with all possible structured keys. This is given by substituting from equations (46), (47) and (48) into equation (24):

$$S_C = 2^{[H(K)-DN]} = 2^{[R'-D]N} = 2^{[R+R'-1]N} \tag{49}$$

Examining the last exponent in equation (49) we see that if the rate R of the source messages, and the rate R' of the channel noise, are such that

$$R + R' < 1 \tag{50}$$

then decrypting the received N-sequence with all structured keys will generate, on average, less than one permitted source codeword by chance alone. Since the true message will — with high probability — also be revealed by this process it will be easily identified.

Thus total reliability — error-free communication — is possible if inequality (50) is satisfied. But since

$$R' \ll 1 \text{ bit per character}$$

for a real digital channel and R is under the communication engineer's control, this condition can always be achieved. Thus the essentially cryptographic technique of soft decision decoding is a valid error correction algorithm.

Now, from our previous definitions, the capacity C of a binary symmetric channel, corrupted with noise of rate R', is given by

$$C = \frac{\max [H(X) - H(X|Y)]}{N} = 1 - R'$$

and inequality (50) may thus be written:

$$R < C$$

Hence the condition for soft decision decoding to give error-free communication over a noisy channel is simply equivalent to operation below the Shannon limit — as it must be.

If soft decision decoding is applied to the wider case of a multi-level channel with L characters, then the condition for reliable communication is a generalisation of inequality (50), namely

$$R + R' < R_0 \tag{51}$$

where, as usual:

$$R_0 = \log_2 L$$

Now, by inequality (32), the condition for a running text one-time pad cipher to be insecure — to be operated beyond its unicity distance — is

$$2R < R_0$$

We see that this is merely a special case of inequality (51), where the source language and the corrupting noise, or key, have the same structure ($R' = R$).

Equivalently we may regard inequality (51) as an extension of the insecurity condition, for a running text one-time pad cipher, to the case where the plaintext source and the enciphering key are messages in different languages.

Again we see that a code operated below its Shannon limit and a cipher operated beyond its unity distance are one and the same thing.

9.13 CRYPTOGRAPHY WITHOUT KEY EXCHANGE

Conventional crypography requires the prior exchange, between the transmitter and the receiver, of a secret key — an exact copy of the randomising noise — by some secure means: usually a courier. The only cipher that has total theoretical security is the random one-time pad. This, therefore, requires the exchange of a key equal to the total length of all the messages that could be transmitted between courier drops. To use such a system on a channel with a high message traffic is thus impractical.

To overcome this problem, new ciphers have been developed that do not need the exchange of secret keys. In such 'public key' systems, an encryption key for use at the transmitter is published — no security is required. A different, but related, key is used for decryption at the receiver. Since only this latter key need be kept secret — and it remains permanently at the receiver — no exchange of secret keys by secure means is required.

Such systems have practical, rather than theoretical, security. They depend not on the ambiguity of the decryption process, but on its difficulty. At the heart of every public key cipher is a computational problem for which no polynomial time algorithm currently exists. Calculation of the secret decryption key from the known encryption key is equivalent to a solution of this problem, and can thus be made to occupy an arbitrarily long period of time. The reverse calculation, however, is computationally trivial and enables the receiver to publish an encryption key without difficulty. Such problems, for which a polynomial time algorithm exists in one direction but not in the other, are known as one-way functions.

In the best known such system — the RSA cipher — encryption involves locking up the message in a modular remainder, while decryption requires knowledge of the two prime factors, p and q, of this modulus. Since no polynomial time factorisation algorithm is known, and p and q are each of the order of one hundred digits long, it is safe to publish the modulus pq for encryption at the transmitter, and still keep the factors p and q secret for decryption at the receiver. As an American number theorist has put it:

'Microdots of p and q are swept up by your cleaning lady and taken by the garbage collector. Their product pq is safe. How to recover p and q? It must be felt as somewhat of a defeat for mathematics that the best strategy by far is to search the garbage dump.'

Public key ciphers, with their reliance on one-way functions and complexity theory, have a totally different philosophy from traditional cryptography. However, under certain conditions, we can use an error correcting code to develop a more conventional cipher that, nevertheless, does not require the prior exchange of a secret key.

Consider, for example, a digital link between A and B (as in Fig. 9.3) having a tapwire to an eavesdropper E. From the nature of operation for such a tapwire, it will have a higher corruption probability — it will be noisier — than the primary channel. We will consider the limiting case in which the link AB is noiseless and the tapwire is noisy and can be considered as a binary symmetric channel with corruption probability p.

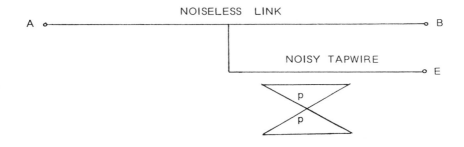

Fig. 9.3 — Digital link with eavesdropper.

All cryptography involves some form of randomisation of the message that can be reversed by the valid receiver, but hopefully not by an invalid one. Conventional ciphers achieve this with a key — a common record of the corrupting noise — and public key systems by the use of one-way functions. In the present example, however, we simply use an error correcting code. This code may be set up quite openly between the valid users, A and B, using the link itself: no prior collaboration is needed.

When A wishes to transmit a message, he sends not the message itself, but any corrupt verions of it that is within the power of the code to correct. In geometrical terms he randomly transmits any N-sequence that lies in the same Hamming cell as the intended message. Since the link is noiseless, a sequence passed between A and B remains in the cell in which it was transmitted. The code itself ensures that only one valid message lies in each cell; thus decoding is unique, and totally reliable communication can be achieved between the valid users. But, as we shall see, total security can be achieved as well.

Since the tapwire is noisy, an intercepted N-sequence does not necessarily lie in the same Hamming cell in which it was transmitted — the eavesdropper experiences irreducible errors. The redundancy of the code — and the degree of corruption it can tolerate and can thus be incorporated — may be tailored to augment the tapwire noise level to the point where the eavesdropper E remains totally ignorant of the true message. The corruption probability q of the extended channel — which includes both the source randomisation and the tapwire noise — can be made virtually total $(q = \frac{1}{2})$.

Randomly transmitting any sequence within the Hamming cell of the message — and none beyond it — is equivalent to making the noiseless link noisy, and then driving it at exactly the Shannon limit. By increasing the redundancy of the code — and hence the noise — this limit can be lowered to the point where E lies, proportionally, so far beyond it that the residual errors corrupt all the information in the message.

For example, suppose there are only two possible messages — M_1 and M_2. These can be coded as N-sequences so that the whole of the Hamming space is tessellated into 'black' and 'white' cells in the manner of a multi-dimensional chess board. Any sequence in the black cells corresponds to the message M_1, and any sequence in the white cells corresponds to the message M_2.

Since there are only two possible messages, each N-sequence carries only one bit of information, smeared out, as it were, along its total length — the rest of the sequence is totally random and uncorrelated to the message. This single bit of message information can just be extracted by the valid users, since an N-sequence retains the colour of the cell in which it was transmitted when passing along the noiseless link. However, N may be made large enough so that, as the sequence passes through the tapwire, sufficient errors accumulate for it to lie, with equal probability, in the black or the white cells when received by the eavesdropper.

Thus the fact that the eavesdropper knows the precise error correcting code being employed — it was set up quite openly using the tapped link itself — is of no avail to him: he cannot extract the necessary information, and remains in total ignorance of the true message.

What is the minimum code redundancy necessary to achieve this situation? An exact mathematical analysis is rather involved, but the following intuitive argument gives essentially the correct answer.

If the rate of the code is R then, since the source is randomised to the limit of this code, the maximum message information that can be extracted from it, even by the valid users, is R bits per character, or NR bits per sequence.

Now considering the tapwire as a binary symmetric channel, with corruption probability p, its capacity C_{TAP} is given by

$$C_{TAP} = 1 - H(p) \tag{52}$$

and the tapwire therefore loses $H(p)$ bits of information per character.

Thus provided the code rate R does not exceed R_{MAX}, where

$$R_{MAX} = H(p) \tag{53}$$

no information remains to be extracted by the eavesdropper.

The minimum code redundancy D_{MIN} is, by definition, related to this maximum rate by

$$D_{\text{MIN}} = 1 - R_{\text{MAX}} \tag{54}$$

and hence substituting from equations (53) and (52) we have

$$D_{\text{MIN}} = C_{\text{TAP}} \tag{55}$$

We arrive at the elegantly simple result that the minimum code redundancy necessary to achieve total security over the link is equal to the channel capacity of the tapwire.

Thus, in this case, we have displayed the use of an error correcting code in a cryptographic mode that achieves both total reliability and total security of communication, without the need for the prior exchange of secret keys. The minimum redundancy required to achieve these simultaneous goals is given by equation (55).

9.14 CONCLUSION

This chapter has explored, in an heuristic manner, the common conceptual basis of coding and cryptography, and has used the binary symmetric channel as a simple model on which to develop these concepts. This model has displayed the theory of codes and ciphers as a coherent whole, enabling an understanding of the one to reinforce an understanding of the other. In particular, error correcting codes were presented as an example — like English — of a structured source language, and decoding algorithms and cryptanalysis as pattern recognition attacks on this structure.

The aim throughout has been to present an overview — no attempt has been made to develop the detailed mathematical techniques of either subject. Indeed no mathematics was used beyond logarithms and elementary probability theory, but still considerable insight was gained from our simple channel model. This was possible because the common link between coding and cryptography — Information Theory — has such far reaching consequences from a few intuitively obvious axioms: a beautiful theory with beautiful results.

Part 3

Networks

Almost all modern communication systems can be viewed as the transmission of data between the nodes of a network. These nodes may be, for example, telephone exchanges, switching centres, computers or satellite stations, and, concomitant with the growing sophistication of the systems, there are often very complex topological arrangements involved. Thus the tasks of trying to ensure that network nodes are not overloaded, that they operate reliably and satisfactorily, and that the routing of data through a network is, in some sense, near optimum can necessitate the use of intricate mathematical techniques.

The first two chapters in this part, by Thompson and Turner, give, respectively, an interesting overview of current and recent work at British Telecom on Telecommunication networks, and an insight into the areas of mathematics that have found useful application in that organisation to the analysis of networks. In the third chapter, Wight describes applications of modelling to the study of the design and configuration of local area networks. This is followed by a contribution from Everett and Tyler which surveys the curriculum content of Higher Education courses with respect to its relevance to the activities involved in constructing 'real' networks (and indeed to the coding task). Finally, in Chapter 14, Dobson makes the observation that standard reliability techniques may be usefully applied to the design of secure distributed systems.

In summary it is clear that, to build satisfactory communications networks, not only is there a need to apply the mathematics of graph theory, but mathematical modelling and simulation also have important roles.

10

Predicting the evolution of communication networks

J. E. Thompson
Network Systems Department, BTRL, Martlesham Heath, Ipswich, UK

10.1 INTRODUCTION

Telecommunication networks are subject to rapid market, regulatory and technological *change*. The rate of change and indeed the *existing* formidable levels of complexity encourage extensive use of computer-based modelling and mathematical analysis to all three facets in order to predict (and hopefully drive) network evolution.

Markets are by nature socio-economic, and are not the concern of this chapter though a mention should be made in passing of a major modelling exercise in which BT is heavily involved which is to define the collaborative European research programme, RACE. Key contributions so far have been to the network reference model which allows various costing and market scenarios to be studied for a possible basis of following technological drive toward the Integrated Broadband Communications (IBC) network.

Regulation is by origin political, and if markets are socio-economic then regulatory effects are either socio *or* economic (cryptic) but either way not very predictable! It may be marginally more predictable for the near term, but the Study announced this year (1987) by the Ministry for Industry and IT into UK Communications Systems Infrastructure is highly significant. The results of this will certainly affect the outcome of the proposed review of Government telecommunications duopoly policy in 1990.

The author's job, on the other hand, is the creation of *technological* change. As a way of structuring this chapter, on how BT develops and uses modelling techniques, it is proposed to refer to five 'enabling technologies' which are having profound influence on the likely evolution of networks.

— *Optical technology* promises virtually unlimited reach to transmission and availability of bandwidth for new services.

— *Radio technology* provides — flexibility (and competition for fixed local links
— network diversity in trunk and international communications
— mobility (cellular and cordless) for personal communications.

— *Switching* is rapidly being *digitalised* for speech and other 64 kbit/s services with the roll-out of Systems X and Y, offering immediate advantages of reliability (more integration), faster call set-up and also promising service flexibility. Note, however, that current design of the call control software provides only point-to-point 64 kbit/s services. We need to plan future 'multiservices' which require complex 'crosspoints' and also control software of such complexity as to require the use of formal methods for computer program creation and evolution.

— *Network topologies* or architectures are, in effect, models which demonstrate more or less flexibility for the incorporation of the above three technologies in evolutionary fashion to meet market demand, an example of which is the Switched Star Network now working in Westminster using optical transmission to provide fully interactive broadband service to 150 000 homes. Modelling can also help predict race or overload states and allow introduction of optimised dynamic networks which generally assist in maximising traffic flow.

— *Artificial intelligence* emerges from the opportunity created by more complex and appropriately organised computing structures to model human behaviour either in the embodiment of Expert Systems of for the design of improved Man–Machine Interfaces which can 'de-skill' the processes of operation, maintenance and control of increasingly complex communication networks.

A number of key examples will indicate how modelling enables at least a partially structured approach to quantifying the potential of these technologies and the degree to which it is possible to predict the technological opportunities for network evolution.

10.2 OPTICAL TECHNOLOGY

Optical transmission has come a long way in an incredibly short time — from patenting of processes to pull fibres of high purity in the 1970s to today where BT already has 105 000 miles of optical fibre installed. World records of increasing bandwidth and diminishing loss have tumbled progressively to current silica fibres exhibiting only 0.2 db/km loss and enabling 100s Mbit/s transmission for 100s km between regeneration. Laying of TAT8 — the first Transatlantic Optical Submarine Cable — will begin later this year — 280 Mbit/s systems with 40–53 km repeater spacing, but already fluoride fibres promise repeater spans approaching 1000 km. Behind all this development lies massive activity to model, predict and analyse performance of fibre systems:

— refractive index profiles through cross-sections of monomode optical fibres can be controlled for minimum dispersion, but this may no longer be necessary, noting the development of narrow-line-width DFB lasers;

— prediction of hydrogen infusion which caused an early TAT8 scare that gas liberated by magnetohydrodynamic action would lead to rapid acceleration of submarine fibre loss — a problem overcome by oxynitride fibre coatings or selection of phosphorus-free glasses;

— optimisation of optical receiver sensitivity by noise models to maximise repeater spacing for a given fibre transmission loss;

— Soliton modelling to predict limits to loss-free propagation of optical pulses in fibre;

— accelerated life testing for ultra-high-reliability integrated circuits and optical regeneration in submarine systems.

10.3 RADIO TECHNOLOGY

Radio provides an alternative transmission medium to optical or coaxial cable in the trunk network, and accounts for some 25% of BT's capacity. Radio-link planning relies entirely on propagation modelling in order to operate services to a specification involving stochastic parameters, an BTRL has been very active in developing microwave frequency planning and spectrum management systems.

10.3.1 PEACEMAKER

PEACEMAKER is a software suite to assist planning and protection of microwave radio networks taking into account the combined effects of the atmosphere and earth's surface on microwave propagation.

The objective is to automate many of the complex processes involved in optimising the 'packing density' of co-channel radio systems. Such systems may be terrestrial or space (satellite), and often share common frequency allocations.

There are many propagation mechanisms which can cause inter-system interference. For example, terrain effects include diffraction and scattering from obstacles; Tropospheric effects include ducting and scattering.

Terrain effects are generally responsible for long-term propagation of radio-waves beyond the visible horizon whereas atmospheric ducting can cause enhanced transhorizon signal levels for relatively short periods of time.

'PEACEMAKER' predicts interference between any two systems (for example), the effect on a proposed new radio link of interface from existing radio networks) by selectivity accessing and manipulating information from databases containing relevant parameters:

Network database — contains the co-ordinates of network stations, antenna heights, etc.

Antenna database — records antenna type and angular gain discrimination patterns;

Terrain database — ground height data from which an interpolated path profile between two defined locations can be drawn (essential in order to calculate terrain diffraction);

Coastline database — enables maps and location data to be displayed;

Radio-meterological database — contains statistical data on the incidence of meterological phenomena which could cause anomalous radio propagation within a geographical zone.

Relevant data automatically selected from the databases provide the inputs to the propagation models within 'PEACEMAKER' which predict interference levels at the required locations.

Propagation modelling is now mature and has long been the subject of international (CCIR) agreement. It is also becoming commercialised, and somewhat excessive confidence in certain quarters is even encouraging change of the Government regulatory and licensing processes towards 'spectrum pricing'. An often quoted and oversimplified analogy is to regard radio spectrum as 'real estate' which can be managed, partitioned and sold or leased for different market applications in different geographical locations. In such a regime, the intellectual property of software suites such as PEACEMAKER will acquire considerable value according to their ability to reliably plan services with minimum probability of mutual interference.

10.3.2   Antenna design
Antenna design is another 'black art' which lends itself well to precision modelling to give the required radiation patterns after reproducible manufacturing processes. For example, GRASP is a program for analysing single or dual reflector antennas of axisymmetric or offset configuration. The reflectors may be defined as having any continuous mathematical form or may even be described by a number of discrete points. A variety of feed options are available, ranging from simple rectangular horns to more complex hybrid mode devices such as corrugated horns. Linear, circular and elliptical polarisations are supported. The scattered fields from the antenna are calculated using an induced current method (Physical Optics), or for wide-angle radiation by an alternative edge diffraction theory (Geometric Theory of Diffraction). The gain of the antenna is also calculated.

Presentation of data is available as arbitrarily defined pattern cuts of both amplitude and phase, or as contour plots.

The program has been used to analyse earth station and terrestrial-link antennas varying in size from 150 mm to 30 m in diameter and has been demonstrated to be very accurate. The program has been further developed to enable synthesis of shaped reflector antennas, an example of which is an antenna to cover an azimuthal sector for multipoint radio.

10.3.3   Computer-Aided Design (CAD)
Computer-Aided Design (CAD) offers a third type of modelling applied to radio.

TeMCAD is a CAD package which has been developed by microwave design engineers at British Telecom Research Laboratories. It incorporates layout and drafting facilities with microwave circuit design functions and produces photo-ready artwork for mask production.

Complex circuits can be built by calling up library cells and interconnecting components. An example shows a low-pass bias filter combined with two tapered sections to form a voltage controlled attenuator. The filter is simply called up from menu by specifying the cut-off frequency and substrate thickness. The tapered sections can be drawn to a selected curve such as a raised cosine.

10.4 SWITCHING TECHNOLOGY

10.4.1 Failure analysis

By the end of 1988 there will be 900 local digital exchanges in operation, providing 2 million customer connections. By mid-1988, the inter-city trunk network will be nearing completion with 55 System X units and 9 5ESS units. There will be enough digital plant in operation to provide a competely digital national network, and as a result, the BT network will be able to run synchronously. However, switching centres will still be connected by line systems with plesiosynchronous multiplexors which were necessary during the digitalisation program to allow exchanges to run independently under their own local 2 Mbit/s clocks. These line systems will then be unnecessarily complex, and failure analysis indicates considerable benefit in replacement by synchronous multiplexing, or 'transwitching'. Such analysis suggests an improvement in MTBF for a typical 100 km, 140 Mbit/s transmission system from 0.29 to 5 years.

10.4.2 Synchronous versus asynchronous broadband local networks

Synchronous multiplexing offers benefits of modular integration and flexibility to broadband local access as well as to trunk planning by simplified multiplex and 'drop-and-insert' techniques. Taking a basic 68×2 Mbit/s 'synchmux' module, we can construct a 'transwitching' network of regular star structure or including synchronous 140 Mbit/s rings with $n\times64$ kbit/s or $n\times2$ Mbit/s access.

However, an alternative view is held, for example, by some participants in the RACE project that better flexibility is offered by ATD networks — high-speed packet switched LANs which offer variable bit-rate services the ultimate 'panacea' for planners of broadband services! Many LAN protocols are unsuitable for real-time speech services on account of their delay — particularly under heavily loaded conditions. BTRL developed the ORWELL protocol, in 1982, with inherent load control mechanisms for mixed traffic allowing prioritising of traffic queues and minimum delay to real-time speech at the expense (for example) of less critical data. ORWELL indeed is an example of the very creation and validation of a concept by modelling alone. Only in 1987, some five years later, are we creating in silicon a hardware realisation of a multiservice switch which will use layered ORWELL rings, or a 'torus' to obtain the necessary capacity, and to demonstrate effective operation of its protocols. One possibility is to realise the Service Access Switch which terminates 140 Mbit/s access via the monomode fibres currently being deployed to major customers.

10.4.3 Multiservices

As mentioned earlier, System X (and ISDN) will offer point-to-point 64 kbit/s switched service. How will we switch video services (more of video *coding* later)? BT currently offers a comparatively limited videoconference service with a multipoint capability allowing simultaneous audiovisual communications between 3 and 5 locations. The author believes this is an essential for conferencing — in a situation where typically 3–10 people from as many locations meet to discuss images and data. Such a substitute for travel will only be acceptable when telecommunications can connect numbers of audiovisual terminals in multipoint configuration, offering an instantaneous mixture of media and a bit-transport system of variable rate and connectivity. How else do we conceive a multiservice, multirate, multipoint 'cross-point' than by ATD? However, the control and management (e.g. billing) software looks horrendous! Even System X (1 million lines of code ?) provides major problems of creating and developing software, and such multiservice control has been estimated as ten times more complex.

Future generations of control software will be far more complex than today's purely by virtue of the demand for ever-increasing systems intelligence, yet we are close to the limits of our ability to design and control such complex systems. The 'engineering' approach is to abstract and model the problem in simpler form: to increase the 'precision' in which we expressed the requirements, the specifications and the design of software, then to 'analyse' the logic of our statement. Of course, for many of us, this involves a wholly new branch of mathematics — discrete mathematics. The next decade will see the move away from our craft-based methods and indeed to a science/mathematical-based discipline. Trends therefore include:

— formal method for complex software structure, i.e. the formalisation of the design process by applying mathematical notations and theorems rather than empirical langues and maxims;
— standard interfaces in software structures — separation of higher layers of control software which will give service differentiation and even allow new services in third-party control software and database.

10.4.4 'Unbundling' and the development of standards for software interfaces

In order to understand these latter concepts of 'unbundling', it might be helpful to present a more complete view of the UK digital network. This will be based on System X in the Digital Main Switching Units (DMSUs) and a combination of System X and System Y (Ericcson AXE 10) and small rural systems known as UXD 5 in local catchment areas called cells. Digital Cell Centre Exchanges will act as tandem switches for the Digital Local Exchanges and Remote Concentrator Units. International connections are via Digital International Switching Centres (DISC) and there will also be a network above the DMSU network of Digital Derived Services Switching Centres (e.g. 0800 services) together with two intelligent network databases (INDBs) to provide other sophisticated services. The System X DMSUs will form a fully interconnected trunk network — connected by the plesiosynchronously multiplexed line systems, mentioned earlier. Such complexity of interconnected stored program controlled switches is driving proposals for fundamental changes to the way in which the control and switch are realised. The major part of the control and management software and data are increasingly expected to reside in a processor system which is quite separate from the switches themselves, these latter using a variety of technologies — synchronous or asynchronous — even optical in future but with only a small spc unit. It will be necessary to conduct extensive modelling studies to determine the validity of the concepts and also to determine the composition of the standard interfaces between the two entities. Common interest by administrations seems likely to demand that these are internationally agreed interfaces. Further modelling will then be required to ensure that any compromise interface eventually agreed is satisfactory.

Standard systems and software interfaces are perceived as important with the advent of competitive software supply in a telecommmunications industry driven by increasing liberalisation. The contribution to understanding given by the seven-layer model of Open Systems Interconnection (OSI) can hardly go unmentioned. Nevertheless, computer communication protocol standards like those defined in OSI have set a precedent for showing how standards based on abstract functional layers can provide a wide range of computer services. There is considerable scope in the future for applying similar techniques to telecommunication services (Open Network Architecture, or ONA).

A comprehensive interface set, associated with a layered abstract model, allows users wishing to deploy new higher-layer services the ability to 'add value to' the lower-layer services. Specification of third-party directory services, for example, can be provided in this way. However, it will require standardisation of functions, primitives and protocols associated with other system perspectives (for example storage) before the full flexibility of the higher-layer services can be fully exploited. This separation of higher-level 'intelligence' from the regular interconnected 'transport' layers has attracted the term 'unbundling' and seems critical to future procurement policies and service differentiation for future network evolution.

10.5 NETWORK TOPOLOGIES OR ARCHITECTURES

Progress in optical, radio and digital technology provides the challenge for BT's network evolution from a current £8.4 Bn dominantly telephony business through an increasingly liberalised environment towards future broadband markets. Many aspects of modelling can be noted. In the local network we are:

— planning Integrated Digital Access or (ISDN) using models of the copper/ aluminium pair network and maximising the opportunity of this considerable 'historic' investment.
— seeking to introduce electronic reconfigurability (ECLON) into the network, providing easier service provision, lower fault liability, centralised maintenance, and consolidation of the existing network 'layers';
— studying the potential convergence of telephony and cable TV and the further prospects of High Definition Television. For example, BT pioneered the switched star network to provide cable TV services to Westminster Cable Company and provide a demonstrator for interactive broadband services. A possible route forward to achieving benefits of convergence of Cable Networks and telephony services is the 'Joint Venture Network'.
— However, broadband has proved to be something of a paradox — noted by the GAP market study commission for RACE — will future interactive networks be broadband or will distribution remain aloof with low-bit-rate codecs providing the interactive visual services?

— Two particular examples relating to architectures, local network and picture
coding are worth treating in more detail, together with a look at the possibilities
of dynamic solutions for optimised traffic models.

10.5.1 Local network model
We have a local network testbed (physical model) from which we are able to derive
parameters for computer modelling of transmission systems. This simulates the
transmission characteristics of the copper pair network and has been used principally
for cost/performance assessment but also to defer construction of costly and time-
consuming physical prototypes until later in a development. In particular, modelling
has allowed:

— simulation of the overall reach and crosstalk improvements available from digital
signal processing techniques such as echo cancellation;
— simulation of the performance of digital transmission linecodes. This has been of
the utmost value in our contribution to North American Standards Committees
on the linecode debate — our models of 2B1Q and the competing systems'
performance have proved consistently accurate;
— derivation of planning rules for the implementation of selected systems in the
field;
— research into the use of machine intelligence to aid local network fault location.

It should be mentioned that one of the principal enabling technologies of digital local
access by such complex techniques is VLSI — CAD provides an underlying example
of the application of sophisticated modelling techniques.

10.5.2 Picture coding
The potential market for cable television distribution (PAL now, MAC, HD-MAC
and HDTV in the future) and new interactive services such as video telephony and
videoconferencing provided the main stimulus for the Westminster Switched Star
System and to justify continued study of broadband transmission in the local
network. Future networks will eventually offer high-speed digital transmission for
which the video codecs will be simpler (and therefore cheaper) than low-bit-rate
video codecs for the same picture quality (although VLSI will reduce the cost gap).
All-digital (high-bit-rate), optical fibre cable television systems could well be viable
within the next decade. Already the marginal costs on a large scale of providing an
optical network are comparable to replacement costs of the much more restricted
copper network, but there needs to be visible commercial incentive to deploying new
network on such a large scale. Incentives are not strong either for new entertainment
TV to residential customers or for interactive visual services to the business
community. In the meantime, beyond Westminster and selected business initiatives
such as the City Fibre Network, the ISDN (2×64 kbit/s) and special private digital
networks (up to 2 Mbit/s) will have to cater for all visual telecommunications needs,
and continuing attention will be given to improving data compression techniques to
maximise picture quality and to minimise the associated digital transmission rate. At
BTRL, a whole range of video codecs is being developed — for broadcast television

distribution at rates from 34 to 140 Mbit/s, for videoconferencing at rates from 64 kbit/s to 2 Mbit/s, and for video telephony at 64 kbit/s. For the last-mentioned in particular, the real technical challenge is to make the integrated codec/terminal cheap enough to be attractive to a business customer on a 'one-per-desk' basis. Still-picture coding schemes for applications such as photographic videotex at 64 kbits/s are also being implemented to reduce the time for a high-quality still picture to be transmitted from about one and a half minutes to a few seconds.

All the moving- and still-picture coding algorithms are investigated and optimised on powerful computer simulation systems. The moving-picture simulation system consists of a VAX750 host processor, a programmable video input/output frame store and a pair of 300 Mbyte Ampex parallel transfer disk drives, and is capable of capturing and displaying 30 s of full-size moving colour pictures in real time. A second system, comprising a Microvax and three programmable video semiconductor memory stores (20 Mbyte in total), is used for processing high-quality still pictures, and is capable of processing and displaying up to 20 s of small, moving colour pictures.

Till now, picture coding algorithmic research has concentrated on exploiting similarity (redundancy) between picture elements, closely adjacent in space and time. Future research will include methods more closely modelled on the processing out in the human eye–brain system. Work on such knowledge-based picture coding schemes is already under way at Martlesham.

10.5.3 Traffic modelling

Switch design has long been the domain of traffic (Erlang) modelling to provide an acceptable grade of service with the minimum number of switch crosspoints and amount of hardware. With the advent of digital switchblocks, which can be economically implemented with negligible blocking, the emphasis turned to the control processors. Much of the analytic work on throughput then required queueing theory while design of the all important overloaded control mechanism required a comprehensive simulation model.

This pre-liberalisation activity from the BPO as the UK PTT contained significant work on behalf of the System X manufacturers but, as a result of the 1981 Act, switching systems are now procured competitively and their analysis by BT has reduced to that of specification and post-delivery check only. The switching system manufacturers have seen a corresponding greater demand for performance analysis to validate their products.

The relief from individual node performance has allowed a greater emphasis in BT on total analysis, which has led to methods of specifying and then achieving a quantified level of end-to-end performance, together with the cost of doing so, thus achieving a better service quality. Again, comprehensive modelling was necessary.

The network studies combined with the opportunity to introduce elegant routing algorithms into the control software of the nodes have created a load position for the UK in dynamic alternative routing. Many administrations have studied and trialled dynamic routing systems, mainly in the form of an overload computer control network. The UK philosophy has been to design a simple algorithm for the software of trunk switching systems so that normal traffic is diverted to alternative routes under localised congestion by autonomous action at that node. The algorithm achieves near optimum distribution of traffic over available resources and its operation can be remotely controlled from zero to full application at any node by adjustment of trunk reservation parameters. (Trunk reservation ensures that fresh traffic is not adversely affected by overflow traffic by reserving a few circuits.) Exceptional overloads focused on one part of the network are handled by an overlay traffic management network using techniques such as call gapping.

Dynamic Alternative Routing exploits or counters such effects as:

— non-coincident busy periods, e.g. business and social traffic;
— outrages caused by faults of accidents;
— traffic fluctuations caused by the introduction of new services (novelty traffic), occasional promotions by customers, large business or social gatherings, etc.

Mobile communications has become a major area of network growth and in a sense provides the ultimate example of dynamic network management. Mobility uses the cellular principle to re-use spectrum so that calls in progress must then be transferred from one radio base site to another as the communicator(s) roams. The handover process can be analysed in terms of the handover traffic which has similarities with overload traffic but must have priority over fresh traffic. Modelling of the many options for initiating handover and processing the event is a new dimension to the highly active traffic modelling which has already occurred to enable the UK cellular mobile service to grow to over 120 000 customers in the two years since it opened in 1985.

10.6   ARTIFICIAL INTELLIGENCE

One of BT's key objectives is to achieve end-to-end management of customers connections as soon as possible. Such extensive network monitoring and control will provide effective 'health care' for the total spread of telecommunications services offered by the company.

End-to-end network management offers:

— up to the minute status information regarding the performance of the network;
— automatic analysis of traffic flows, giving advanced information of possible congestion;
— collection and analysis of network fault reports;
— automatic diagnosis of network failure;
— control of technicians;
— build-level information.

The intention is to integrate the network management functions with the administrative procedures concerned with service provision and thereby create an efficient and cost-effective solution to the provision of telecommunications. Such an integrated approach is creating an extensive and complex Telecommunications Management Network. In turn, Artificial Intelligence (AI) will be the key to practical management of this complexity.

— AI techniques are being used to provide improved man–machine interfaces in order that network management operators can comprehend the total situation.
— Expert Systems are being developed which will assist with the location of faults and service restoration.

AI draws expertise from a broad range of interests including robotics and neural networks, speech and image undertanding.

An example of BT's use of AI is to assist with fault analysis of modern SPC (System X) exchanges. In conventional expert systems the knowledge concerning the application domain is encoded in a set of production rules. In condition . . . then action.

Rule-based expert systems are satisfactory for many domains, but in engineering systems, the numbers of rules would become prohibitively large. Fortunately, the structured nature of the domain has allowed more powerful knowledge represen- tation techniques to be developed. Essentially these techniques model the hierarchi- cal structure of exchanges in a way which permits the program to draw logical inferences on the impact of faults on the overall performance. In general, therefore, the Expert System has three elements — knowledge acquisition, an inference 'engine' (often in the form of existing or proprietary shells) and an explanation facility which informs the user how the program arrive at any particular conclusion.

Current research is concentrating on the use of the Object Orientated paradigm to encode the exchange model itself. Logic languages such as Prolog have been used with some success, but these suffer from run-time problems such as large memory requirements. Such 'second generation' Expert Systems are particularly relevant to telecommunication networks since they are highly structured and their orderly construction readily permits the use of modelling techniques.

A simple example of modelling addresses a classic problem in managing an extensive telecommunications network — how to get the correct technician to the site of a problem: the 'dispatcher problem'.

Current research at BTRL has found a solution which uses three models and logical inference:

— The geographical area is modelled: roads, location of exchange, etc.
— The equipment (exchanges, etc.) is modelled.
— The technician's skill and work load are modelled.

The inference mechanism is able to schedule a fault to a technician having regard to:

— The location of the fault and the technician (Route finding).
— The priority of the fault (Problem ranking).
— The skill and expertise of the technician (simply, Human modelling).

This system is now being field trialled by BT's operational units. Full deployment of such a system will, of course depend on staff acceptance of such techniques.

In summary, AI will find application in several areas — intelligent front end to existing (complex) hardware and software, graphics and natural language interfaces and increasing use of Expert Systems.

10.7 CONCLUSION

The author feels that this chapter lacks connectivity and apologises for that. However, even though the use of modelling is now quite pervasive in telecommunications research, the particular applications and models *are* themselves disjoint. It is hoped that the author has given some idea of the range and types of activity at Martlesham, and one particular area will be further developed by Turner (Chapter 11). Expertise *is* specialised but there is common interest — also the IT era is bringing convergence between those in telecommunications and those in the computer forum to encourage the dissemination of modelling methodology. The POLYMODEL conference makes a welcome initiative in the North East of England.

To conclude, one hopes to be profound. What is certain through the difficulties of predicting such rapid technological change is that future networks will be literally incredibly complex — at the component level, the system level, the service level and the operator interface level. Modelling will be essential to design, test, install and manage such networks. If the technical models are sound, we can nervertheless define interfaces and standards and we can attack the markets with confidence.

The author by referring to RACE — probably the biggest modelling exercise ever mounted — and driven by the will to survive of the entire European electronics industry. BT is leading Europe as a privatised administration, and at the end of the day, the market model is the business case. Nothing of any substance — INDB, Satellites, Submarine Systems, etc. — gets bought without a sound business case to support the investment decision. The BT network has strengths which lie in its weaknesses. The Local Network is both its existing means of deploying the ISDN and its greatest fault liability. The company has a fixed asset value of £10½ Bn and is spending on modernisation at about £2 Bn a year. It also employs about 230 000 people, hence the need for the implementation of AI to maximise operating efficiency as networks get ever more complex. It is with this need for commercial justification of network evolution in mind that the author has picked certain key examples of modelling and simulation. In these areas, technology potentially will drive markets, and the models will provide essential inputs to the business case for the deployment of that technology.

11

A survey of recent network studies

P. M. D. Turner
Performance Engineering Division, British Telecom, Martlesham Heath, Ipswich, UK

11.1   INTRODUCTION

This chapter gives an overview of the research and development work carried out by the division in support of the network planning departments. This work includes telephone network analysis and call routing studies.

The main concern of these studies is the congestion performance or blocking of the network under consideration. In this context we usually concern ourselves with link blocking, although models of processor (exchange) blocking are occasionally used.

This work generally comes under the following headings:

—Evaluation of routing strategies
—Dimensioning
—Performance evaluation
—Network management
—Signalling network

Each section starts with the description of a technique followed by the description of an application of that technique. Where possible, an example is given of the methods used. In order to do this, a great deal of complicating detail has been omitted. The most important and relevant work areas are described below in sections 11.2, 11.3 and 11.4. Section 11.5 describes an interesting technique that we have already found useful in an optimisation problem. Section 11.6 is included as a short mention of other work undertaken and in particular that on cellular mobile radio.

11.1.1   Some terms and the Erlang-B function
To aid clarity, we list a few terms and their usage:

Blocking	— when a call fails owing to the lack of a resource.
Grade of Service	— often abbreviated to gos, the probability that a call is lost.
Node	— general term for telephone exchange or switching centre.

Traffic stream	— the total traffic originating at a particular node and destined for another individual node.
Link	— the set of traffic-carrying circuits between two nodes.
Path, route	— a set of links used in carrying traffic from an originating exchange to a destination exchange.
erlang	— a measure of traffic intensity that is most easily understood as the mean number of calls that would be carried on an infinite trunk group.
Erlang-B function	— often written $b = E(N,A)$, this gives the probability of blocking for a full availability (i.e. all calls have access to any free circuit) group of N circuits offered Poisson traffic of intensity A erlangs. Classical analysis assumes negative exponential call holding times, but the function holds more generally.

11.1.2 Routing methodologies

It is not proposed to give a detailed description of all the routing options available in the new digital network. However, it is necessary to explain a few simple concepts before proceeding.

Direct routing	— Only one possible route is possible between the two end points and this is usually the shortest (minimum number of links) path. For trunk calls this would be hierarchical and involve at least three links. In the case of a fully interconnected network this would be a one-link path.
AAR	— Automatic Alternative Routing. This refers to the automatic selection of an alternative if the first choice (the only one available with direct routing) is blocked on the link out of the originating exchange. Blocking on any later link results in a lost call. On a fully interconnected network, AAR would typically be used to set up two link calls via a tandem if the direct link is blocked. There may be one or more possible alternative routings.
ARR	— Automatic Re-Routing. This is an extension of AAR in which a call is not immediately lost if blocked on a link subsequent to the first. With ARR, control is passed back to the preceding node and any remaining alternatives can then be tried.
Overflow traffic	— Traffic on a link can be split into two categories: (1) directly routed traffic (sometimes known as fresh traffic); (2) overflow traffic, which is using the link in one of its alternative routings (due to the action of AAR or ARR).
Trunk reservation	— This is a service protection mechanism available on the System X exchange. It works by limiting the access of overflow traffic to circuits on an alternative path. Use of the

path is only allowed when at least a fixed number of circuits are free on the alternative outgoing link. This number is called the trunk reservation factor. Studies have shown that this method works well at all traffic levels—this is not true of commonly used alternatives.

11.2   ONE-MOMENT MODELS OF LINK AND STREAM BLOCKING IN TELEPHONE NETWORKS

The performance analysis of telephone networks is greatly complicated by the non-Poisson nature of the overflow traffic from a group of circuits offered Poisson traffic. Indeed, no fully satisfactory, general solution has been found to this modelling problem. Under certain circumstances, useful results may be obtained by ignoring this non-Poissonality. Furthermore, the resulting simplification makes it possible to incorporate a number of other factors into the network model. In the following subsection we describe a simple one-moment model of some generality. A number of different one-moment models have been developed by the division, but this should suffice to give a good idea of the methodology.

11.2.1   A simple one-moment method

To explain some of the concepts involved, first consider a simple three-link network:

$$
\begin{array}{ccc}
P & Q \\
V & V \\
T\longrightarrow X \qquad X & X & X \longrightarrow T+P+Q \\
B_1 & B_2 & B_3
\end{array}
$$

where P, Q, T are offered traffics and B_1, B_2, and B_3 are the link blockings. Taking account of thinning and baulking, the traffic A_2 offered to the second link becomes

$$A_2 = T(1-B_1)\,(1-B_3)+R(1-B_3)$$

Here, the $(1-B_1)$ term represents **thinning**, while the $(1-B_3)$ term represents **baulking**.

Since the offered traffics depend on the blockings, an iterative approach is required to determine blockings. For the simple case of streams with disjoint alternative paths and no trunk reservation, the following set of non-linear equations in the link blockings would be solved. This is sometimes called the Erlang fixed-point method

$$B_L = E\left[N_L,\ \sum_{\substack{(i,j,k)\\ \text{s.t. } R_{ijk}=L}} A_i\,\Pi_{n<j}(1-\Pi_p(1-B_{R_{inp}}))\,\Pi_{m\neq k}(1-B_{R_{ijm}})\right]$$

where B_i is the blocking probability of link i,
   N_i is the number of circuits comprising link i,
   A_i is the traffic offered by stream i, and
   R is the routing matrix (R_{ijk} is the kth link of the jth choice path for stream i).

Stream blockings are given by

$$S_i = \Pi_j \, (1 - \Pi_k(1 - B_{R_{ijk}}))$$

In general, the equations would not be so simple.

Results given by this type of model need to be validated by, and checked for accuracy with, simulation.

A very general fixed-point model has been developed as a design aid to the local part of the inland communications digital network. It can be applied to networks of almost arbitrary topology and routings. It can be used in both a fixed-point mode and a distribution mode. The former analyses the performance of the network at a fixed point in time, thus using the fixed-point model once only. In the distribution mode, we analyse the performance of the cell over the entire design period. To do this, all known effects are considered.

— Network size and structure
— Routing schemes
— Trunk reservation
— Traffic variability
— Correlation of traffic streams
— Growth and forecasting error
— Overloads
— Failures

Having decided on the first three factors above, the other factors are taken into account in determining an offered traffic distribution for every stream in the cell. Even for the smallest of networks it would be impractical to analyse the network for every possible set of values from the stream distributions. For this reason, a Monte Carlo technique was adopted. Here, a traffic value is selected from each of the stream offered traffic distributions and the corresponding stream grades of service determined. This process is repeated a large number of times and the results assembled into a probability distribution for each stream. An overall (traffic weighted) distribution can then be formed. These results give information on the end-to-end grade of service experienced by a customer in the local cell.

11.2.2 Network behaviour under overload conditions

Another one-moment fixed-point model has been developed to investigate network behaviour under focused overload and failure conditions. The network modelled includes both the local and the main trunk networks, but is greatly simplified by assuming both a symmetric structure and a symmetric traffic pattern. The only routing allowed is the simple, direct one, i.e. no alternatives are allowed. Two further effects are modelled here: *exchange congestion* and *repeat attempts*.

Exchange congestion

Each exchange is modelled as a single server (processor), finite queue with a constant service rate unless the queue is full when the service rate is adjusted to take into account the time taken to reject calls.

We have

$$\alpha' = \alpha(1 - \lambda\tau)$$

where
- α is the service rate when the queue is not full.
- α' is the service rate when the queue is full.
- λ is the call arrival rate.
- τ is the time to reject a call.

Repeat attempts

Sometimes known as Customer Persistence, this is modelled by assuming that each call attempt that fails will try again with a probability p. Thus one intent to make a call may generate a number of call attempts ending either in a successful call set up or in the customer 'giving up'. Clearly the number of call attempts generated by one intent is geometrically distributed with parameter p. We can show this succinctly in a diagram below.

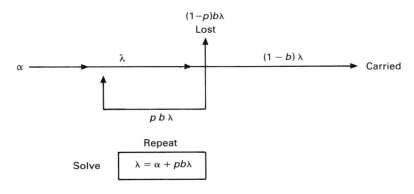

Here,
α is the call intent rate,
λ is the call attempt rate,
b is the stream blocking,
p is the persistence rate.

The results and conclusions from this work are unfortunately of too complex and technical a nature to be reported on briefly here. However, it can be noted that a great deal of insight was obtained into the effects of varying levels of overload on network congestion.

11.3   THE IMPLIED COST MODEL

This model was developed by Kelly [1] as part of a wider ranging collaboration. Considerable effort is being expended by the division to develop it into a useful tool for Network Management and Planning. When a link i is used by a call, there is an expected loss of revenue due to the increased network blocking caused to other calls attempting to use that link. Clearly it is not desirable to set up a call on a path if the total loss of revenue due to the use of the component links is greater than the revenue gained by the call. This notional cost c_i has another interpretation as a **shadow cost**.

This shadow cost measures the increase in the total traffic carried by the network if the capacity of the link is increased by one unit.

Here we make the usual one-moment model assumptions to derive the following expressions.

Let,

$\lambda = \{\lambda_r\}$ be the vector of carried traffics,

$\mathbf{v} = \{\mathbf{v}_r\}$ be the vector of offered traffics,

and

$\mathbf{C} = \{C_i\}$ be the vector of link capacities.

Here, r ranges over routes, and i ranges over links.

By the standard fixed-point one-moment methods we can solve $\lambda\,(\mathbf{v};\,\mathbf{C})$.

RATE OF RETURN (per unit time) $\mathbf{W}(\mathbf{v};\,\mathbf{C}) = \Sigma\,\lambda_r\,(\mathbf{v};\,\mathbf{C})w_r$

where w_r = return per erlang (route r).

SHADOW COSTS are given by

$$c_i(\mathbf{v};\,\mathbf{C}) = \frac{\partial \mathbf{W}}{\partial C_i}(\mathbf{v};\,\mathbf{C})$$

Note the use of lower case c_i for shadow costs and upper case C_i for link capacities.

To reiterate,
$\{c_i\}$ give the increase in rate of return when the capacity on link i is increased by one unit, **but also** appear as the costs of increasing traffic on a route.

11.3.1   Applications
This model has considerable potential, and its use is being investigated in the following areas.

1.   *Network dimensioning*
This involves techniques based on the idea of 'growing' the network adding circuits at each stage by comparing the link cost:return (i.e. shadow cost) ratio for each link and augmenting the most favourable at each stage of an iterative process. At present it seems likely that a method which both adds and subtracts circuits would be needed to produce good results.

2.   *Routing strategy*
Since shadow costs can also be interpreted as the cost of allowing one extra call on a route they can be used to determine which routings give the greatest return. Indeed some routings may be discovered to give a negative return, indicating that they should not be used at all in normal circumstances. This clearly has potential uses in the short, medium, and long term: short term, if it is developed into a routing

algorithm; medium term, if it is used to advise routing changes with changing traffic patterns; long term, if developed into part of an algorithm for determining routings and link capacities (dimensioning) at the same time. The medium- and short-term uses are of interest to Network Managers; the long-term uses would involve Network Planners.

3. Network evolution

The very nature of the algorithms makes it suitable for short-term network evolution. The best link augmentation strategy and routing modifications can be determined at any point during the network's evolution over time.

11.4 DYNAMIC ALTERNATIVE ROUTING

The divisional collaboration with Cambridge University has also been instrumental in the proposal and study of a distributed control routing method called DAR (**D**ynamic **A**lternative **R**outing) [2]. This routing method is proposed for the fully interconnected digital main network.

    If we assume that at any chosen point in time each source–destination pair of nodes has a tandem node in current use, then the method for handling a call can be described in the following steps.

1. If the direct link is free, use it.
Otherwise
2. Try the two-link path via the appropriate tandem for the source–destination pair. If this two-link path is free, use it.
Otherwise
3. Fail the call and reselect the current tandem node for the source–destination pair.

11.4.1 Modelling and behaviour

This method has the following properties.

(a) It tends towards equalising lost traffic on the different routes used by a traffic stream. Thus the proportion of stream traffic sent on route r is inversely proportional to the blocking on route r.
(b) It needs the use of trunk reservation to prevent network instability and protect stream grades of service.
(c) It works best for small overflow in relation to fresh traffic.
(d) It can be modelled (and has been) by a fixed-point one-moment model.

11.4.2 Advantages of DAR

(a) The distributed nature of the algorithm makes it both practical and relatively easy to implement. This is in sharp contrast to many other adaptive routing strategies.
(b) Studies have shown the method to adapt well to changing traffic patterns.
(c) The cost savings in terms of providing a network to a given set of performance criteria can be significant.

11.5 PROBABILISTIC HILL-CLIMBING

Also known as simulated annealing, this technique can be used in all manner of optimisation problems (see [3] for a useful description of this methodology and the properties of Probabilistic Hill-climbing Algorithms).

There are a number of things that have to be done to set up the process.

(a) A *Markov Chain* structure must be imposed on the 'state space' of possible solutions.
(b) A stopping condition is required. This may just be a restriction on allowable computational time.
(c) An *objective* (or cost) function is required.
(d) The probability of accepting a transition to an inferior state is determined by a *temperature* variable. Management of this variable requires a way to determine two things: firstly, the number of iterations at one temperature and secondly, a *temperature* reduction algorithm for which a constant multiplicative factor is often used.
(e) A function of *temperature* and increase in cost that determines the probability of accepting an inferior state. The function $\exp(-c/T)$ is generally used.

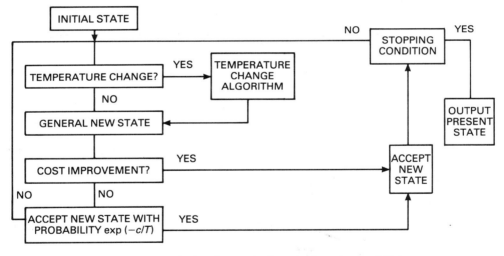

In the above flow chart, c is the change in the cost function and T the temperature.

This methodology has been applied to the problem of determining the optimum AAR routing strategy for the DDSN (Digital Derived Services Network). This is a fully interconnected network with eight nodes and thus 56 traffic streams. For this purpose the network configuration (state) was determined by the tandem selected for each stream. In this case the tandem used between two nodes was chosen to be the same for both traffic directions, thus requiring 28 choices of tandem. The configuration space was clearly too large for an exhaustive analysis.

Let each configuration be represented by a vector $\mathbf{V} = \{v_1, v_2, \ldots, v_{28}\}$. The *Markovian* structure is imposed by defining a configuration vector $\mathbf{U}$ to be a neighbour of $\mathbf{V}$ if and only if it differs from $\mathbf{V}$ in only one dimension. Thus there are exactly 140 neighbours to each configuration, and all transitions take place between neighbouring states. Transition probabilities are defined to be all equal to 1/140. An initial configuration was determined by a simple heuristic based on link sizes and spare capacity. The cost function was based on the mean overall grade of service for the network. An upper limit was also imposed on the worst stream grade of service. Results were a very significant improvement on the initial configuration.

11.6 OTHER WORK

Work has also been undertaken on common channel signalling networks, processor dimensioning, and cellular mobile radio systems. The following problems in the last of these areas have been considered.

(1) Signalling channel capacity. The signalling channel uses a variant of the CSMA/CD random access protocol in which only the monitoring stations can detect collisions. Some very interesting results concerning the stability of this network were obtained [4]. At high traffic levels it is possible for a small increase in calling rate to result in the network flipping into very serious congestion. The network could remain in this state until a large drop in calling rate occurs. This 'hysteresis' behaviour makes adequate capacity planning essential.
(2) Standard voice traffic capacity in the presence of handovers and queueing. A number of different channel allocation algorithms were studied assuming fixed channel allocation within a cell.
(3) Dynamic channel allocation and optimal control.
(4) Possibilities of sectorisation and cell overlap.

11.7 CONCLUSION

We have shown a brief outline of our work on circuit-switched networks. It has become apparent that some themes repeat themselves across several of our large projects. In particular, one-moment models have been extensively used to model some quite complicated networks. Other methods (for example, Interrupted Poisson Processes [5] and Equivalent Random Theory [6]) based on 2 or 3 moments can be more accurate for small and simple networks, but they are not easily extendible to large networks or complex routing schemes. There is still quite frequent recourse to methods involving the theory of stochastic processes, in particular Markov chains, birth and death processes, and queueing theory. However, the main effort has been moving into large network studies where various algorithms have been devised, implemented, and evaluated, computationally. Here it is the very tractability of one-moment models that makes them a useful practical tool.

REFERENCES

[1] F. P. Kelly, 'Blocking and routing in circuit-switched networks', *Int. seminar on Teletraffic Analysis and Computer Evaluation*, (1986) Amsterdam.
[2] R. R. Stacey & D. J. Songhurst, 'Dynamic alternative routing in the British Telecom trunk network', *Int. Switching Symp.*, (1987) Phoenix, Arizona.

[3] F. Romeo & A. S. Sangiovanni-Vincentelli, 'Probabilistic Hill climbing algorithms: properties and applications', *Chapel Hill Conference on VLSI*, (1985).

[4] N. Macfadyen & D. Everitt, 'Teletraffic problems in cellular mobile radio systems', *Eleventh Int. Teletraffic Cong.*, (1985) Kyoto.

[5] A. Kuczura, 'Queues with mixed renewal and Poisson inputs', *Bell Systems Tech. J.*, **51**, No. 6, 1305–1325 (1972).

[6] R. I. Wilkinson, 'Theories of toll traffic engineering in the U.S.A.', *Bell systems Tech. J.*, **35**, No. 2, 421–541 (1956).

12

Modelling workloads for local area network performance analysis

A. S. Wight
Department of Computer Science, University of Edinburgh, UK

12.1 INTRODUCTION

Workload characterisation has been an important consideration in any study analysing the performance of a computer system. This has been true for batch and interactive systems, and studies have been carried out by:

— measuring existing systems in service
— discrete-event simulation models
— analytic solution of queueing network models.

The studies hav been done as part of:

— the design of systems
— tuning of existing systems
— predicting the performance of systems in capacity planning exercises.

There is obviously a very wide range of workload models required for studies which range from detailed consideration of machine instruction sequences to the level of interactive user behaviour and machine independent functional workload models. In its most detailed and complete form, the construction of a workload model consists of data capture from working systems and data reduction, e.g. cluster analysis.

The performance of a computer system is very dependent on the dynamic interaction of various components of the workload. A good workload model will capture these interactions and allow study of the critical aspects of system performance. The difficulty of constructing such models is illustrated by the preponderance of workload characterisation papers which deal only with batch systems and models which contain the right amount of each workload component but fail to control or generate the right mixes.

Compute manufacturers have a particular interest in good workload models. They design, implement, test and market computer systems with particular users in mind and endeavour to have available the best possible, detailed workload models at all stages of development. However, once machines are out in the marketplace they may be used by user groups the manufacturer has not envisaged such as engineering systems used in educational or financial establishments. The manufacturer must therefore develop workload models for these new areas and track their development so that new systems are competitive and the appropriate technology can be employed through the lifetime of these systems.

12.2 COMPUTER NETWORKS

Increasingly, computers are connected by communication links into networks. So we see part of the workload of a computer system coming from its network connections and also that the load on the network is generated by the computer systems attached to it. Experience with attempting to understand computer systems indicates that results have to be analysed and presented for each class of user. The average response time for an average user simply does not convey enough useful information. If we are going to model a network how do we characterise the workload and avoid falling into the same trap of oversimplification? A simple, perhaps biased, view of network modelling suggests that

● models are generally simple
● there are few reports of measuremens of working systems to enable detailed validation of models
● modelling studies have concentrated on hardware/communications media and associated protocol levels.

What we need are

● models to deal with current workloads rather than very simple, idealised models dictated by solution techniques
● models to deal with changing workloads
● the ability to deal with higher-level protocol issues, such as policies and flow control
● consideration of the effects of technology changes, e.g. buffering in interface modules.

In many system studies, the workload model (benchmark) is fixed to allow comparisons for management/political/economic reasons. This is also true in the case of manufacturers mentioned above. In fact, users adapt and change with new technology, both hardware and software, and in response to the service they get.

Likewise an existing LAN workload model may have to be changed because of user and technology changes which cause parameter values for the workload to need adjustment, e.g. packet interarrival times may be influenced by software and hardware overheads in the node/adaptor/station interface. A new or changed interface may affect the workload seen by the LAN for the same application. Buffering determined by a protocol may fix packet size and this will influence the characteristics of the workload.

The kind of model we use may depend on the type of study we are doing. There is a difference between studying design trade-offs and doing detailed capacity planning.

● Simple workload models may be good enough to expose significant differences in performance between different designs, particularly studying the performance of individual hardware components in great detail.
● Good capacity planning to take account of workload changes and growth requires much more detailed knowledge of the workload. In this case we are now dealing with absolute rather than relative performance.

Experience with driving multi-user systems via a Remote Terminal Emulator has indicated that experimental runs modelling large numbers of users can be very good for debugging and stress-testing of the system but it is dangerous if the numerical results are taken out of context because it is too easily forgotten that in this environment 'users' are out-of-date models and not real, current users. The influence of user behaviour on a time-sharing system is similar to that of the load offered by a node to a Local Area Network.

12.3   LOCAL AREA NETWORK PERFORMANCE MODELLING

Local Area Network performance is going to crop up as an important issue in a wide range of expanding areas of computer and communication systems such as

● distributed operating systems
● database management
● transaction processing on-line systems
● resource scheduling
● graphics
● performance analysis tools.

Better measurement and modelling will certainly be required as network loads increase to the point where network performance is an important component of the response seen by users and applications. For example, integrated voice and data traffic is going to generate a range of new considerations. Data packets generally have a bimodal size distribution. There are many more small packets, but they only contribute 20% of the bytes transferred. Voice traffic requires a bounded delay on transit time. For data, variable delays may be acceptable; but, for bulk data, some minimum throughput is desirable.

12.4 NETWORK GROWTH

As networks of machines become larger, the nature of communications is changing from stream-orientated (file transfer) to transaction communication (page access and Remote Procedure Call), i.e. request–response. Can we get by just welding existing components together or do we need to redesign network interfaces and transport protocols to take account of

- high-speed communications media
- mismatch in speeds between nodes on the network
- file transfers, which must be done successfully in a request–response environment, that is, delay must be controlled if there is a user/application waiting.

In particular, as transmission speeds rise, users are encouraged to consider applications where large files require to be transmitted. This has encouraged investigation of protocols which send bursts of packets without waiting for acknowledgement of individual packets. This high-speed transmission can lead to congestion and subsequent packet loss because the receiving nodes cannot cope. Packets must therefore be re-transmitted, e.g. complete file, from first detected loss, selective re-transmissions of lost packets only. This type of problem is particularly likely to arise where a bridge links two high-speed networks.

Controlling the flow of packets in a network successfully at lower levels may simply mean that the delays have been shifted to become problems at higher levels, that is packets are waiting outside the network rather than in queues in the network. The aim must be to avoid packet re-transmission caused by packet loss or timers being triggered. Possibilities are to increase the capacity of the network, make greater use of alternative routing and prevent users starting if they are likely to cause trouble.

As computer communication networks have grown over the years, the area with most serious problems has shifted between processors and communications networks and back again. Moving to ever higher-speed networks has put problems at the nodes, that is the interface between processors and networks. Because memory is cheap, some of the problems can be smoothed by improved buffering. Nevertheless the discussion above leads us to ask how big the buffer can be? If large files are being moved and the protocol allows the possibility of complete re-transmission, should we attempt to keep the complete file around until safe delivery is acknowledged or is the cost of retrieving it once more from backing store acceptable?

12.5 SUMMARY

In conclusion, workload modelling of networks must start at and take greater cognisance of the user level of activity. The influence of higher levels of protocols must be incorporated since they are going to play a greater part in determining performance than the characteristics of the transmission medium where so much past modelling effort has been concentrated.

13

Network and coding courses at undergraduate and diploma level

M. G. Everett and **F. M. Tyler**
Thames Polytechnic,

13.1 INTRODUCTION

The recent increase in the use of electronic communications has naturally led to an increased demand for specialist knowledge in codes, ciphers and networks. The interdisciplinary nature of these subjects has meant that courses have evolved in computing, engineering, and mathematics departments. In this chapter we examine courses at undergraduate and Higher National Diploma level; whilst we pay particular attention to the progammes of studies in mathematics departments, comparisons are made with courses in engineering and computing.

It should be noted that simple ideas of codes, ciphers and networks have been included in engineering courses for some time. The more sophisticated techniques which had been developed were of no use to the practical engineer since the processors required to implement the ideas were too slow and expensive. The micro-revolution suddenly changed the position, and it is now not only possible but cheap and desirable to include the more advanced methods in practical communication systems. It is also the case that there are many additional application areas, for example digital recordings, video-phones, which use concepts and techniques from codes, ciphers and networks.

Finally, mathematics department have taken an interest, particularly in coding theory because it allows esentially pure mathematicians to offer 'applied' courses. There seems to be increasing pressure from students, industry and government organisations on institutions in higher education to provide vocationally orientated graduates. Certain areas of codes, ciphers and networks contain highly developed concepts from abstract algebra, and these have consequently attracted the pure mathematicians looking to put an applied slant on their courses.

The major part of our survey was to look at courses in mathematics departments in codes, ciphers and networks. The *British Combinatorial Bulletin* lists 20 institutions which run courses in the area of investigation. All the courses at undergraduate level were final-year options, with coding theory being taught at nearly all these colleges. The courses were usually one-semester courses, and this has been taken as the basic unit length.

13.2   CODING THEORY

The majority of the courses at universities have as a prerequisite a solid foundation in algebra. It is expected that the student is familiar with subjects such as finite fields, rings, Galois theory, etc. All the courses contain a core of material, namely linear codes, generating matrices, Hamming codes and cyclic codes (e.g. BCH). In addition the courses contain further material which differs from institution to institution. Some departments decide to look at more advanced work such as Golay codes or the relationship of coding to Steiner systems. Others look at information theory, considering such topics as entropy, channel capacity and Shannon's theorem. Not unexpectedly the criterion for selection of material is mathematical interest.

In the polytechnics, a similar pattern emerges, except that there is no assumption that the student has studied the necessary abstract algebra. Hence, typically, courses consist of finite fields, rings, linear codes, generating matrices, Hamming codes and cyclic codes. In addition, some courses look at further codes such as BCH whereas others consider practical aspects such as choice of code or television image reconstruction. Courses from engineering departments (only a few were examined) contain similar material on the surface. All obviously contain Hamming codes and cyclic codes. However, the emphasis is, not surprisingly, very different. Much less attention is paid to the theory and much more attention is paid to the results. Also consideration is given to the encoding and decoding devices, material not usually covered in the mathematics department courses. One of the most interesting differences is the inclusion of convolution codes in some syllabuses. Mathematicians have little interest in these codes mainly because the theory is not as well developed and is far less elegant. A natural consequence of this is the exclusion of this material from their courses. It should be noted that convolution codes are extremely important. It is these codes that NASA uses for its deep-space programme. Coding theorists would probably claim that a course in coding theory should contain a large proportion (close to 50%) of material on convolution codes.

13.3   CIPHERS

Courses on ciphers seem to be rarer; there was only one course completely devoted to cipher systems in our survey and this is at Royal Holloway and Bedford New College, University of London. Elsewhere, ciphers are usually taught as an integral part of a larger course (usually coding theory). The reasons for this are probably three-fold. Firstly, it is a newer subject (at least in terms of published material on open access). Secondly, applications are not as common as codes. However, there has been a recent upsurge in interest in cryptology. This interest is due to new legislation and the fact that companies now require safe communication systems together with secure data storage. Finally, as far as mathematics courses are concerned, the mathematically interesting techniques are not always very practical, and the practical techniques are not particularly interesting. For example, stream ciphers are mathematically interesting but do not provide very secure systems, whereas DES which is widely used is not mathematically very deep.

The course at RHBNC consists of early cipher systems, stream and block ciphers, speech security and public key systems. The cryptological part of the other courses is usually a distillation of this material, with emphasis on the more mathematically interesting areas. Hence the material would consist of substitution ciphers, ideas of block and stream ciphers (usually including a sketch of DES) and public key systems (particularly RSA).

Again there is a distinction between the engineering approach and the mathematical approach. Engineers describe safe systems, will probably not prove any results, but do, however, look at more practical problems. It should be noted that this includes not only the physical implementation but also the important area of management of secure systems.

Both approaches have their difficulties: the engineer will probably not appreciate the security inherent in a particular systems and will find it hard to assess any implications of advances in cryptography. On the other hand, the mathematician could easily miss an important consideration in managing a system and, since any system will only be as strong as its weakest link, totally undermine its security.

An interdisciplinary approach is obviously required. This approach should not be thought of as the combination of an engineer and a mathematician. Each individual will have specialities which do not necessarily merge together to form a coherent approach. For example, an engineer and a mathematician working together on an RSA system might easily come up with the idea of using a pseudo-number generator to create large primes. A cryptologist would realise that this approach is not possible for an RSA system, since an attack could be made via the prime number generator. An interdisciplinary approach requires a cryptologist who understands both the engineering and the mathematical aspects of the subject.

13.4 NETWORKS

The major difficulty in surveying network courses is that the word conjures up a different subject to different groups of people. There are in essence three distinct types of network course. Firstly, there is the operations research approach. This is typified by the course at Southampton University, where the major topics are event networks, topological ordering, resource allocation, Ford–Fulkerson, matchings and multi-commodity flow. Although many of these topics can be applied to problems in communication networks, this is not an overriding factor in the selection of the material. In contrast, the second type of course concentrates on computer networks, and is usually found in computer science or communications degrees. The emphasis is not so much on the structure of the network but on how the machines interact. A good example of this type of course is the one at Salford University which covers such subjects as networking in commerce, public domain networking, e.g. Kermit, protocols and local area networks. In the third category are the mathematical courses which develop graph theory for applications in computing and communications. Most universities and an increasing number of polytechnics teach introductory graph theory — some with applications. The approach most relevant to communications beyond introductory material is typified by the course at Thames Polytechnic. Subjects covered include connectivity, planarity, networks with applications, e.g. Polish notation, minimal connector problems, shortest-path algorithms, Ford–Fulkerson, backtracking and NP-completeness. Courses of the above type are relatively rare, and most mathematical graph theory courses still pursue the more traditional approach to the subject. The diverse nature of network courses makes it more or less impossible to make any general comments. It is of interest to note that the mathematical side of networks, i.e. graph theory, is receiving more attention in many colleages and that this interest has undoubtedly been fuelled by applications in information technology.

13.5   HIGHER NATIONAL DIPLOMA COURSES

Although courses at undergraduate level are relatively common, our survey only uncovered one course in a mathematics department of Higher National Diploma level. It seems likely as the popularity of such courses increases that more will filter into the lower-level courses. The one course found was entitled Mathematics for Information Technology and included some material such as Boolean algebra which was outside the scope of our survey. Since it is impossible to draw conclusions from a sample (or population) of one, we record the contents of the course for interest: Group theory; Codes — group codes, generating matrices, Hamming codes; Applied graph theory and networks–connectivity, planarity, Eulerian tails, maximal shift register sequences (using Good diagrams), Polish notation (via arborescences), Ford algorithm; Boolean algebra; cipher systems — substitution ciphers, block ciphers (affine), DES and RSA.

13.6   CONCLUSIONS

The success of the courses above depends on the objectives behind their construction. If we look at student opinions, they seem very favourable. It must be remembered that all the courses above were options and it is therefore likely that students would enjoy a topic they have selected. Alternatively we could look at whether the courses prepared the students for employment. In the areas outlined above there is a growing demand for the more talented students and many are now obtaining jobs in relevant areas. If we take a broader view, it is often the case that the student who opts for courses in communications will also be doing other IT options. This blend of material will make them a desirable commodity in the jobs' market, and the shortage of suitably qualified people is likely to remain with us for some time.

Mathematical modelling for communications is obviously a useful source of interesting material for inclusion in courses at a variety of levels. Our plea is that such courses adequately reflect the truly interdisciplinary nature of the subject. There is a real danger that courses in this area will go the same way as certain applied mathematics courses which examine such real problems as frictionless surfaces, incompressible laminar flow and semi-infinite domains. Mathematical modelling was a natural reaction to courses for which the mathematics came first and the problems followed. We ask that modelling in IT does not start from real problems and disappear in the direction of vector spaces over arbitrary fields never to return to the practitioner again.

14

Reliability and security issues in distributed systems

J. E. Dobson
Computing Laboratory, University of Newcastle upon Tyne, Newcastle upon Tyne, NE1 7RU, UK

14.1   INTRODUCTION

In this chapter we briefly examine links between the two apparently distinct topics of security and reliability, and argue, as has been implied by Laprie [1], that they can usefully be regarded merely as different aspects of a common problem, and so susceptible, at least in part, to common solutions. In so doing, we challenge some of the current approaches to the design of highly secure computing systems, and suggest instead that an approach which has fault tolerance as its basis can achieve a high degree of security as well as reliability.

A recent paper by Randell [2] argues that a correct approach to systems design involves treating distinct issues, including reliability and security, as logically separate problems, provisions for which should be made by logically separate mechanisms. It is the thesis of this chapter that, at least as concepts, reliability and security are not necessarily best treated so separately, and that their joint consideration can lead to some interesting new insights.

Some links between the subject of reliability and that of security are of course obvious, and well-acknowledged. Most notably, one would expect that due effort is spent ensuring adequate reliability of the security-critical mechanisms in any systems which purported to meet any significant security criteria. Also, one would surely normally expect information which is extremely secret also to be extremely valuable, and so require it to be held reliably, as well as securely. Quite how this is done without the reliability mechanisms compromising the security mechanisms is another matter! Indeed, it was such concerns that prompted the stress on separability of security and reliability mechanisms in the earlier paper. However, we believe that the two topics have much more in common than merely being linked as requirements, and the remainder of this chapter will explore these more interesting similarities before briefly discussing the implications for future standards.

14.2 DISTRIBUTED SYSTEMS

The approach to the design of highly secure systems that we are suggesting is very deliberately taking as its starting point an environment containing so-called 'distributed computing systems'. By this term we mean a system made up of multiple computers, interacting via a non-instantaneous communications medium, capable of working (and failing) independently of each other. Such computing systems might be spread over a large geographical area, or, in the not-too-distant future, might co-exist on a single silicon wafer or even chip. But the problems of security and reliability have a wider domain than mere computers, and our real interest is in 'distributed systems', i.e. in systems which could, for example, include people and machines that are interacting with, and perhaps through, one or more computing subsystems. Against such realities the problems of an isolated computing system look very specialised and uninteresting, and even treatment of distributed computing subsystems, without considering the human environment in which they are placed, can sometimes engender a feeling that what is being addressed is only a part of the real problem. Nevertheless, it is convenient to start by considering just the distributed computing component of a system.

The relevance of distributed computing systems to reliability is well known. Though they pose additional reliability problems, they also provide the basis for a variety of reliability mechanisms. The essence of these mechanisms is redundancy and separation — the provision of additional separate means of storing, communicating and/or processing information, so that errors can be detected and recovered from, or masked, so that failures can be averted. However, as Rushby [3] and others have argued, 'separation' is also at the heart of the security problem — and manifest physical separation is one of the simplest and most easily verifiable forms of separation.

It is possible to try to build complex software and hardware mechanisms which will guarantee that a given single computer sytem prevents, say, top secret information leaking into unclassified files. However, one can, we believe, more easily achieve an equivalent effect by assigning the separate component computer systems of a distributed computing system to separate security regimes, in the knowledge that the only security-critical mechanisms will then be those relatively simple and isolatable ones that are involved with inter-computer communication.

This is what has been done, at the Royal Signals and Radar Establishment in the UK, where a prototype distributed multi-level secure system has recently been constructed [4]. The system uses UNIX United, and is based on the design which was originally outlined in a paper by Rushby and Randell [5]. Oversimplifying somewhat, each separate UNIX system looks like a directory in the overall system, and runs at a single security level. The rules concerning permissible information flows between levels are enforced by placing the trusted mechanisms for enforcing security policy in specially designed network interface units, which are also responsible for appropriately encrypting all inter-machine information communication. As a result, no trust has to be placed in any of the individual UNIX systems [6]. Moreover, the overall multi-level secure system still appears to its users to be a conventional UNIX system. Thus on practical, as well as philosophical, grounds we regard security as a problem which is best addressed within the framework of distributed systems.

14.3 RELIABILITY

During recent years, research groups at Newcastle and elsewhere have spent much time seeking improved definitions for the various fundamental reliability concepts. Our own dissatisfaction with the hitherto typical definitions of terms such as 'failure', 'error' and 'fault' arose early on, when we started to consider the possibility of providing what is now termed 'design fault tolerance', particularly for software. We found it inappropriate to start from an enumerated list of possible faults, which had been the hardware engineers' approach — this makes little sense when one wishes to allow for the possibility that design faults, of unknown form, are still lurking somewhere in the deeper recesses of the system. Instead we took the notion of a 'failure' i.e. the event of a system deviating from its specified behaviour as our starting point, and then defined 'reliability', 'fault' and 'error' in terms of 'failure' [7].

A brief recapitulation is in order. The occurrence of a failure must be due to the presence of defects within the system. Such defects will be referred to as *faults* when they are internal to a component or the design and *errors* when the state of the whole system is defective. Even though the external state of a component may be an error in the system of which it is a part, the component need not be in an erroneous state when it is considered as a system in its own right. The internal state of the component may be perfectly valid but incompatible with the states of other components of the system. Thus the only difference between a fault and an error is with respect to the structure of the system, and the distinction between error and failure is that between a condition (or state) and an event. We can then say that a fault is the cause of an error (i.e. transition to an erroneous state) and an error is the cause of a failure (i.e the event of not producing behaviour as specified).

With such an approach, if one has more than one specification for a single system, each capturing some different requirement which is to be placed on its operation, one has thereby defined different types of failure, and indeed different types of reliability. However, to avoid unnecessary confusion, we will follow the lead given by Laprie and use the term 'dependability', to encompass the different types of 'reliability'. This will enable us to adhere to the more common informal usage of reliability as relating to functionality and the continuity of its achievement.

The principal techniques for trying to achieve reliability can be classified into fault prevention and fault tolerance — often regarded as rival, but more profitably as complementary, techniques. Fault prevention attemps to ensure that an operational system is fault-free, either because any faults it contained were removed before it was put into service, or by avoiding the inclusion of faults from the outset. In contrast, the fault tolerance approach accepts the possibility that, despite whatever fault prevention efforts have been made, faults might nevertheless still be present in the operational system. In particular, fault tolerance is a way of handling residual design faults, and fault tolerance techniques such as design diversity are now becoming common in safety-critical computer applications, though more frequently for purposes of error detection than fault masking. The first level approach to this situation is to construct system components using masking redundancy (e.g. N-Modular Redundancy for operational faults in hardware, and N-Version programming for design faults in software) and to assume that all faults are indeed successfully masked.

A further degree of sophistication involves allowing for the possibility that such fault masking may not always be successful (or even appropriate in some cases), and equally that a given component might in any case sometimes be invoked incorrectly. Such considerations have led to the development of exception handling schemes which have the property that normal and abnormal states are clearly defined and differentiated, that the error detection and recovery mechanisms are well structured, that exception interfaces are as well defined as the normal ones, and that the overall scheme can be recursively applied [8]. Such schemes are therefore of potential applicability at all levels of system design, but they do depend crucially on the application of a uniform exception handling model throughout the whole system and are much more dependent on a coherent error recovery strategy than they are on the sophistication of the error detection mechanisms.

14.4 SECURITY

It is very noticeable to us that work on secure computing systems has typically concentrated on the use of (security) fault prevention techniques. For example, penetration exercises have been used as the basis of a fault removal strategy. However, the preferred approach is to try to avoid faults, by insisting on formal or semi-formal analysis of the specification at each level, and verification that it is met by the design and its implementation. But this concentration on fault prevention, and in particular on fault avoidance, seems to apply only to the computer component of a secure environment. In a wider environment, one would expect to find plans and protocols for recognising and dealing with security leakages. After all, capturing and turning spies, for example, is in effect a form of fault location and error compensation, i.e. of fault tolerance in practice. It is perhaps surprising that more attention has not been paid to fault tolerance techniques when it is required to achieve security in computer-based systems.

However, before the designer can assess the possible applicability of fault tolerance techniques to the problem of ensuring the security of a computing system, it is necessary to consider the role of a secure computing system as a component within a secure enterprise, alongside the other components such as sentries, hidden cameras, barbed wire, or whatever. One expression of the requirement for its introduction might be that the computer system introduces no new security flaws into the overall system, i.e that it is itself intrinsically secure. There are two criticisms that can be made of such a requirement. The first is that it is unrealistic to believe that it is possible, just as it would be unrealistic to believe that the introduction of a new and complex component into a system would not alter the reliability characteristics or would not introduce any new failure modes into the system; and the second criticism is the system structuring one that the whole system should have a recursive reliability or security failure model with well-defined exception interfaces for the signalling of failures and that **all** components should conform to this model.

It should be noted that the comments that we have been making apply not only to the construction of a would-be secure computing system out of a set of (hardware and software) components, but also to the whole system, viewed as a component in some larger environment. Thus it is always prudent to doubt that the system is in fact as secure as its designers and certifiers allege, and also to allow for the possibility that accidental or deliberate actions by users might cause system behaviour that, though 'correct' with respect to the security specification, is nevertheless later found out to have been inappropriate. When these comments are taken into account, it can be seen that one task of the secure system designer is to consider what can be done entirely within the computing system, and what aids can be given to the people in the environment of the system to assist in the forward and/or backward error recovery procedures necessary to limit damage following a security violation. (The problem is similar to that of assisting database system users who find out that their database has for some time contained, and has been giving them, incorrect data [9].)

14.5 SUMMARY

Our work on fault tolerance for achieving system reliability has led us to understand that the definition of reliability solely in terms of internal behaviour of a component is inadequate. As we explained earlier, one has to consider overall system state as well as the state of a particular component, and thus specify the behaviour of the system as well as of each part of it. Such specifications will characterise the desirable behaviour of the component at its interfaces with its environment, and also define any means it is designed to have of signalling such exceptions as are admitted might occur. (In principle, any such specification should aim to be as complete as possible — in practice, one might choose to abbreviate the specification by making it explicitly dependent on the security specifications of the sub-components and by assuming that these are not violated [10]. Every effort should be made to devise and incorporate into the system cost-effective run-time checks against possible failures to meet these specifications, as well as provisions for responding to indications that externally applied error checks have revealed security violations. Such internal and external checks should supplement any replication and majority voting schemes which are used in the system. Moreover they should be fitted into a carefully thought out, and fully general, exception handling scheme.

In summary, we are recommending that security be viewed just as a special case of dependability, and be the subject of similar, if not identical, design approaches and techniques to those used for achieving high reliability. Thus we are saying that one should attempt to employ a methodology based on building secure systems out of insecure components, or more accurately, less insecure systems out of more insecure components. One should not rely totally on security fault prevention techniques, but rather one should use an appropriate blend of security fault prevention and security fault tolerance.

ACKNOWLEDGEMENTS

This chapter is a somewhat shortened combination of two earlier papers written by the present author in conjunction with Brian Randell [11,12]. The author is grateful to Brian Randell for his permission to take sole responsibility for the current cocktail. Considerable benefit, as well as enjoyment, has been gained from discussions with Tom Anderson, Derek Barnes, Mike Martin, and Dan Nessett. The work on reliability and security at the University of Newcastle upon Tyne is sponsored by the UK Science and Engineering Research Council and the Royal Signals and Radar Establishment, Malvern.

REFERENCES

[1] J.-C. Laprie, 'Dependable computing and fault tolerance', *15th IEEE International Conference on Fault-tolerant Computing*, pp. 2–11, June (1985) Ann Arbor, Michigan.

[2] B. Randell, 'Recursively structured distributed computing systems in *Proceedings of the 3rd IEEE Symposium on Reliability in Distributed Software and Database Systems*, pp. 3–11, October (1983).

[3] J. M. Rushby, 'The design and verification of secure systems', *ACM Operating Systems Review*, **15**, No. 5, December (1981).

[4] J. Wood & D. H Barnes, 'A practical distributed secure system', *Proceedings of the International Conference on System Security*, pp. 49–60, October (1985) Online, London.

[5] J. M. Rushby & B. Randell, 'A distributed secure system', *IEEE Computer*, **16**, No. 7, 55–67, July (1983).

[6] K. Thompson, 'Reflections on trusting trust', *Comm. ACM*, **27**, No. 8, 761–763, August (1984).

[7] P. M. Melliar-Smith & B. Randell, 'Software reliability: the role of programmed exception handling', in *Proc. Conf. on Language Design for Reliable Software*, Raleigh, NC (*Sigplan Notices*, **12**, No. 3), pp. 95–100, March (1977).

[8] T. Anderson & P. A. Lee, *Fault Tolerance*: *Principles and Practice*, (1981) Prentice–Hall, Englewood Cliff, NJ.

[9] S. K. Shrivastava, 'A dependency commitment and recovery model for atomic actions', in *Proceedings of the 2nd Symposium on Reliability in Distributed Software and Database Systems*, pp. 112–119, July (1982) Pittsburg.

[10] R. B. Neely & J. W. Freeman, 'Structuring systems for formal verification', in *Proceedings of the 1985 Symposium on Security and Privacy*, pp. 2–13, (1985).

[11] J. E. Dobson & B. Randell, 'Building reliable secure computing systems out of unreliable insecure components', *Proceedings of the 1986 IEEE Symposium on Security and Privacy*, pp. 187–193, April (1986) Oakland, California.

[12] B. Randell & J. E. Dobson, 'Reliability and security issues in distributed computing systems', *Proceedings of the 5th IEEE Symposium on Reliability in Distributed Software and Database Systems*, pp. 113–118, January (1986) Los Angeles, California.

Part 4

The social impact of communication

There can be no doubt that the changes in communication technologies, some of which have been described in the first three parts of this volume, are having far reaching effects on social, economic and political organisations in third world, as well as advanced, countries. It is therefore entirely relevant that a discussion about the communications industry should not neglect the implications for society of these new technologies.

In the first chapter in this part, Melody describes the ESRC research programme which he has been co-ordinating. We too often, faced with an information glut, lose sight of the human dimension to advances in Information Technology, to the social and political fact that technical innovation must be assessed and applied. Thus co-ordinated research teams are being funded to study various facets of the impact on society of advances in Information Technology. In the second chapter, Robins and Hepworth attack the common failure to realise that technology is always social. They argue that it is too often treated as if it was a neutral phenomenon that promised uninhibited progress and does not belong in the ebb and flow of industrial history and politics. According to Robins and Hepworth there is a continuity between old and new industrial ages and a failure to appreciate this increase the probability, identified in Melody's chapter, that the already advantaged gain disproportionately. They believe that the economic patterns of new technology in the UK are now becoming clearly traceable and bear this out.

In the final chapter, Hacker and Monge point to the deficiencies of the Shannon--Weaver model. They think that the fault lies in a failure to do justice to the human and social factors. Social processes are involved in both coding and decoding, and humans are not passive in these activities. These authors believe that a more complex model, which takes account of the interactive and recursive processes of communication, is required. They suggest that a failure to produce this more complex model may account for the low use rates experienced in the USA for apparently major breakthroughs, and that there is a confusion between Information Theory and Communication Theory.

15

UK research on implications of information and communication technologies†

William H. Melody
Director of Programme on Information and Communication Technology,
Economic and Social Research Council, London, UK

15.1 INTRODUCTION

Technologically advanced countries have grown increasingly dependent upon electronic information and communications systems. Most of their institutions have been structured around these systems. Explanations of the functions and effects of these institutions are premised upon (often implicit) assumptions about the state of underlying information and communication networks. When these networks undergo major change, many explanations of economic and social phenomena developed from the study of the old order are reduced in significance or rendered obsolete. Uncertainty is increased. The new information and communication systems are often more complex than the old. Understanding is more difficult to obtain, yet more important to pursue. Ironically, in an age where information and communication systems are more sophisticated and comprehensive than ever before, the planning horizons for decision makers of all kinds are continuously being reduced because of a growing inability to forecast even short-term future developments. Seldom in our history has a subject attracted such attention, yet yielded so little critical insight into its long-term implications.

The growing significance of electronic information and communication networks has brought to the foreground the recognition of the overwhelming importance of information and communication, aspects of life that heretofore generally have been taken for granted. The characteristics of information generation and dissemination effect the nature of markets and the structure of industry, as well as the competitive-

† This chapter first appeared as an article in *Telecommunications Policy*, **11**, Number 1, March (1987), (Butterworth Scientific Limited, PO Box 63, Westbury House, Bury Street, Guildford, GU2 5BH, UK), and the editors are grateful to Colin Blackman, Managing Editor of *Telecommunications Policy*, for giving permission for the article to be included here.

ness of firms and the prosperity of regions. They affect the internal structure of organisations ranging from corporations to government agencies, political parties, universities, trade unions, libraries and volunteer groups. They affect the formation of social and cultural networks; the characteristics of work and leisure; the role and content of formal education; the structure, content and effective control of the mass media; and the information environment through which public opinion is formed.

The implications of the changes now taking place in the information and communication sector are made all-pervasive because they affect the characteristics of essential information and communication networks both for individuals and for organisations. Yet most theoretical models attempting to explain the characteristics of economic markets, political democracy, or the cohesiveness of social and cultural groups tend to assume a stable state of perfect information and communication. This may be tolerable under relatively stable conditions. It is not tolerable when the information and communication structures are undergoing significant change.

15.2   THE NEED FOR RESEARCH

Information and communication issues are becoming more and more central to public policy in all nations and in a growing number of international agencies [1]. Because of its pervasive impact upon society, the information and communication sector is not easily separated from other sectors. Essentially it consists of microelectronics; computer hardware, software and services; telecommunication equipment and services; the mass media and a plethora of new database and information services, as well as the more traditional forms of information and communication such as print and postal service. Stimulated by rapid and continuing technological change, this sector has experienced a high rate of economic growth in recent years. Moreover, the direct economic effects are compounded by the fact that major parts of this sector provide important infrastructure services or enabling functions that affect the operation and efficiency of almost all other industries, as well as government agencies and most other institutions. Information gathering, processing, storage and transmission over efficient telecommunication networks is the foundation upon which technologically advanced nations will close the twentieth century as so-called 'information economies', or 'information societies', i.e. societies that have become dependent upon complex electronic information and communication networks and which allocate a major portion of their resources to information and communication activities.

This sector may have become even more significant to the development of national and international economic development than any of the major transport expansion eras of the past, including canal, rail or highway. It already has become a central concern of domestic industrial policy for most developed countries. In the UK, for example, the privatisation of British Telecom, the creation of Oftel and the cable authority, the Alvey, Hunt and Peacock Commissions, and the adoption of preliminary policies relating to cable television, direct broadcast satellites, invisible exports and training for computer literacy are but a few recent illustrations.

Yet the implications go much beyond the domestic scene. The expansion of the information and communication sector serves to integrate the domestic economy more easily into the international economy by means of efficient international information and communication networks. As international economic integration is expanded, the impact of domestic public policies is reduced. Control over the domestic economy by national governments is weakened. These developments are forcing governments to recognise the need for a full range of international trade policies addressed not only to direct trade in information and communication equipment and services, but also to the implications of world-wide information and communication networks for other industries. For example, these considerations are central to current discussions at GATT relating to the treatment of service industries in international trade.

The international banking and finance industries have already restructured their organisation and methods of operation in the light of the enhanced opportunities for transferring money and data instantaneously around the world. This has been a major influencing factor prompting deregulation of the banking industry in the UK. Many transnational corporations have been able to improve their organisational efficiency and control by centralising more decisions at their world headquarters, while maintaining flexibility in decentralised production. This has raised the possibility that significant decision-making power, as well as research and development and information services' activities, will be removed from 'national' subsidiaries that in some cases have been reduced to the status of branch plants.

Medical, tax, credit and professional information relating to citizens and institutions of one country is being stored with increasing frequency in another. This raises important policy questions in a number of areas, including, for example, the terms and conditions of access to information, privacy of personal information and the scope and limitations of national and regional sovereignty. It raises questions of the vulnerability of a country's economic and political decision-making systems to losses of essential information because of breakdowns in crucial information and communication networks that occur outside the country.

Significant changes in information and communication networks require a reinterpretation of traditional notions of public information (e.g. news, libraries, government reports and statistics), and private information (e.g. strategic corporate forecasts), and the terms and conditions for access to it. In more and more circumstances, information itself is becoming a marketable commodity. There are now thousands of databases in the world selling a variety of information to clients over modern telecommunication networks, and the number is growing rapidly. Proposed changes in copyright laws now under discussion in several countries would permit a further expansion by strengthening legal property rights to information.

15.3   TOWARDS EFFICIENT INFORMATION MARKETS

Continuing growth in the information and communication sector is opening opportunities for growth in a wide variety of information and communication markets, trading in both public and private information. Although these markets are adding value to information, they are very imperfect markets. They raise important questions both of government regulation of monopoly power in national and international markets, and of government policy with respect to access by the public to traditional types of public information. There is undoubtedly a lucrative market for selling news about important national and world developments to special interests before making it available to the public. And the Government could earn substantial revenue by selling advance notice of its money supply statistics or its tax policies at monopoly market prices. Clearly the determination of the appropriate limits to place on the commoditisation of information must be a subject of indepth research, public debate and the crafting of informed public policy.

These developments raise many important issues for social science research. The establishment of efficient markets in 'tradable information' requires government decisions on a number of fronts [2]. Because most saleable information reflects the adding of value to some form of existing information raw material, entrepreneurs must have reasonable access to this raw material. Much of it is held by government agencies and public bodies that traditionally have not sold their information services to the public at market value. There are many artificial restrictions to access that need to be removed.

Indeed some information raw material can only be gathered by public bodies. Therefore, private markets in tradable information may be stimulated by increased activity by public agencies. In addition, although it is important to ensure that there are enforceable property rights for tradable information, those rights should provide an incentive to create and distribute new forms of information, not to protect monopolies of information. Finding the appropriate balance is a difficult task that should be informed by research on a continuing basis.

The successful development of information markets requires major adaptations by both private and public institutions. If markets in tradable information are going to work efficiently and equitably, they must be developed upon a foundation of public information that provides the education and training necessary for citizens to function effectively as workers, managers, consumers and responsible citizens. Determining the appropriate adaptations, both by the public and by the private sectors, to the new information and communication environment is a crucial task to which social science research can contribute significantly.

Many people and organisations can benefit substantially from the rapid expansion of the information and communication sector, but at least some are likely to be disadvantaged, in both relative and absolute terms, especially if traditional public and social services are displaced, downgraded or made more expensive. To illustrate, a considerable portion of the information now accessible through public libraries is subject to commoditisation and sale in private markets, where it would be accessible only through telecommunication-based information services. In recent years, many libraries have expanded access to a variety of bibliographic databases. But they have cut back their physical holdings of government reports and statistics, general research reports and studies, periodicals and even books. This has facilitated research projects with the funding support to pay for computer searches and acquisition of the desired material. But most academic researchers, students, and the lay public can rarely afford to use computer searches, and are increasingly frustrated by the more limited access to hardcopy resources.

The telephone system is being upgraded to the technical standards of an integrated services digital network (ISDN) that is more efficient for the plethora of new information services required by sophisticated high-volume users. But it may be significantly more expensive for small-volume users, and users with only local telephone service requirements. This could make it more difficult to extend basic telephone service coverage to a larger proportion of the population [3,4]. The UK legal database, Eurolex, was recently sold. UK users can now purchase access to electronic UK legal information at a higher price from the Lexis database in the USA.

A major challenge for public policy will be to find methods to ensure that developments in the information and communication sector do not exacerbate class divisions in society and that its benefits are spread across all classes. This will require new conceptions and operational definitions of the 'public interest' and of public services, new interpretations of the requirements of social policy, and the design of new structures for its efficient and effective implementation.

From the perspective of public policy, these changes provide an environment of continuing transition for the foreseeable future. There is an opportunity, and a large responsibility, for policy makers not only to guide and manage the transition in the best interest of society at large, but also to establish a policy framework that incorporates the contributions of all sectors of society in defining the kind of information society that is desired. This only can be achieved if a substantial research effort is mounted to provide the information necessary to generate policy options, evaluate policy alternatives and take informed policy decisions. There is a real risk that the transition to the information society will not be guided by informed policies because of inadequate research and debate on these and other important issues of public policy.

15.4   TRANSDISCIPLINARY RESEARCH

Information and communication developments have tended to erode heretofore separable areas of public policy, and to increase the probability of unforeseen implications arising in areas outside the purview of traditional policy analysis. To illustrate, industrial policy attempting to promote the rapid development of telecommunication facility networks in the UK — for example, broadband cable, direct broadcast satellite and integrated services digital networks — could substantially expand imports of equipment to construct the facilities. the new facility networks may open the country to imports of data processing and information-related services serving a broad spectrum of UK industry and professions, displacing domestic production and services ranging from specialised databases to television production. This, in turn, could create a serious balance-of-payments problem, result in more jobs lost than created, and undercut domestic cultural policy.

A coordinated set of policies, based upon an assessment of the implications of a range of alternative policy combinations, could be developed so that the new technologies would be implemented at a pace, in a direction, and under conditions where the benefits to society are likely to be maximised and the losses minimised. Obviously no 'policy set' will permit governments to exercise total control over the pace and direction of implementation of new information and communication technologies. But a significant degree of influence is possible. Government policies do promote, restrict and influence the direction of change in many ways. And government policies clearly must deal with the implications of changes emanating from the information and communication technologies. In this sense, overall government policy influencing the course of events is reflected in the net effect of the aggregation of individual policies in all related areas, including those where policies of *laisser-faire* and benign neglect are employed. The great challenge for policy research is explaining the complex set of inter-relations between and among policy areas that were previously thought to be reasonably discrete and separable.

Some nations undoubtedly will find it in their interest to promote the implementation of information and communication technologies at a faster pace than others. Depending upon the implications for the domestic economy, other national priorities and the particular values of each nation, the optimal pace and structure of implementation will differ, in some cases substantially. The USA and Japan see enormous benefits from international trade in information and communication equipment and services for their domestic economies. Most developing countries, with extremely low rates of telephone penetration, many urgent domestic priorities and large trade deficits, are much less enthusiastic. For countries like Canada, the UK, and other European countries, the implications are mixed. Structural implications are likely to be significant and policy development more complex.

The difficulty to date has been in understanding the important dimensions of information and communication policy issues, particularly when it comes to assessing the long-term implications. There has been a tendency for governments to recognise only those immediate issues that have been thrust before them, generally in fragmented fashion, outside either a long-term or a systemic context. It is the latter dimension of information and communication developments that provides the challenge for research that can inform policy decisions.

This research is directed toward assessing the long-term implications of institutional change, and therefore requires a focus on particular institutions and all aspects of their relations to other institutions. It is dynamic because both the internal structure of organisations and their external relationships are changing significantly and not necessarily in any uniform pattern. The research must be transdisciplinary because the integrative character of information and communication technologies requires integrated policy research, and the implications cut across many disciplines, including both the social and the physical sciences. The existing arbitrary division of professional knowledge in the social sciences may be useful for examining a stable structure of institutional relations. But it is inadequate to the task of understanding dynamic change of the significance and complexity portended here. Attempts to force information and communication issues into the narrow constraints of the received theory of any social science usually overlook key dimensions of the issues, assume away others, and often lead to misleading conclusions.

15.5   THE ESRC PICT NETWORK

In 1983 the UK launched its Alvey Programme, a massive five-year research and development effort directed at scientific and technological improvements in information technology, primarily computing and its applications. The emphasis is on invention and technological development. This parallels similar initiatives in the USA, Japan and other European countries, as well as European-wide initiatives such as ESPRIT, EUREKA and RACE. The objective of the Alvey Programme is to enhance the competitiveness of UK industry by providing technological advantages of the precompetitive stage in industries that must compete in global markets.

The Alvey Programme the 'T' of 'IT'. It does not address issues relating to the implementation of information technologies in a real economic, social, political and cultural environment, and their implications. It does not address the fundamental UK problem, captured imaginatively by a headline in *The Economist* several years ago: 'Invented in Europe, patented in America and made in Japan' [5]. It does not address the economic, social, managerial and policy issues that must be understood if the potential benefits of the new technologies are to be converted into jobs, wealth, economic growth and human and social well-being. Although on an infinitely more modest scale of funding, the ESRC PICT is directed to partially filling this gap.

Up to six major research centres will be established in the PICT network for an initial period of eight years. Each centre will specialise in certain aspects of the social science and policy dimensions of the information and communication technologies, although each centre will bring research strength that cuts acrosss several disciplines, and therefore have areas of commonality with other centres. The centres will coordinate their research activities so as to obtain maximum benefits within the context of the overall research programme. In addition, these major research centres in turn will provide documentation and related services to the research community at large. The ESRC PICT office will provide overall coordination and direction.

Although the research programme was initiated, and is being directed by the ESRC, it is being developed as a joint programme with the cooperation of the University Grants Committee and the host institutions, as well as industry and government active in the information and communication technologies field. Total expenditure for the programme is expected to grow to about £2 million per annum at maturity. Representatives from participating government agencies and industry are active in the formulation of the research projects, and will parcticipate at several stages throughout the conduct of the research projects.

During the first year, the PICT undertook a substantive review of past and current research in the field and provided network management and service functions to the research community. Contacts with government agencies and industries were established, including a constructive working relationship with the Confederation of Information and Communications Industries (CICI), The Department of Trade and Industry (DTI), the Research and Development Division of the British Library, British Telecom and other organisations. European and other international contacts have been established with institutions that have similar interest in this field.

An important element in the PICT is the training of a cadre of men and women able to research and/or work in the information and communication technologies field. Although there are several postgraduate degree courses in the UK at present, and several more in the planning stage, covering various aspects of the information and communication technologies, an examination of the content of these courses has revealed that, by themselves, they are unlikely to meet the perceived need. It is anticipated that several, although not necessarily all, of the research centres selected for the PICT network will be interested in developing new doctoral programmes in the field. Interdisciplinary doctoral programmes, in association with research centres, should provide a constructive environment for achieving an expansion in research/training opportunities.

15.6   THE RESEARCH CENTRES

Neither policy problems nor research issues on the information and communication technologies are easily structured. The implications of these technologies tend to be pervasive, permeating to a greater or lesser degree the bounds of traditional disciplines, problems and research areas. Yet some ordered structure of the field must be developed as a basis for analysis and ultimately the selection of research programmes and centres. Based on a review of the research literature, the range of important long-term policy issues being raised, and the plethora of researchable problems being generated, the ESRC took a decision that the area in greatest need of immediate attention is 'Implications of the Changing Characteristics of Information in Society', with particular reference to the growing significance of tradable information.

The information and communication technology developments offer major opportunities, significant problems and some potentially serious risks. The issues are not well understood. The research base is extremely small and the area is not well established in research institutions. The need for informed policy guidance is great. Therefore, the first research centres in the network have been selected to address the evolving characteristics of information, with particular reference to policy options and their implications.

The overriding theme guiding the research programmes of the PICT research centres is the examination of the multifaceted role of information and communication in the UK economy and its relation to the global economy, with particular reference to the implications of the information and communication technologies for economic development and for changes in public and private institutions. The research will examine: the characteristics of information markets and the information and communication industries; the effects of laws, policies and regulations on the operating effectiveness of information markets; the implications of new information technologies, telecommunication networks and electronic information services upon the operations of specific industries and the delivery of public services; the changing relationship between public and private information; and the implications of public policy options on matters affected by the information economy. The research results are intended to inform public discussion and to enrich the foundation upon which major government and industry policy decisions are made.

This area of research goes well beyond the resources and competence of any single research centre. Thus, centres were selected which together can provide thorough coverage of this area, while also making research contributions in related areas.

— The Science Policy Research Unit (SPRU) at the University of Sussex has been doing policy-relevant research in areas closely related to the information and communication technologies for many years, including studies of the processes of technical change, patents and innovation, labour markets and industrial policy. It has also made major research contributions on the development of the service sector, including social innovations and information services. Building on its past work, SPRU will focus particular attention upon the empirical basis for assessing the information economy, the implications for industrial competitiveness from IT expenditures in the military sector, the emergence of information and value-added network service (VANS) markets and the implications of software development on economic growth.

— The Centre for Urban and Regional Development Studies (CURDS) at the University of Newcastle upon Tyne has established a strong tradition of empirical work, including many case studies in the fields closely related to the information and communication technologies, that have emphasised the geographic, spatial and distributional implications of economic development and public policy. CURDS has developed substantial data and expertise in labour markets within both UK and European perspectives, including the changing characteristics of education and skill requirements. CURDS will focus its attention on the implications of the information and communication technologies for the geographic distribution of economic opportunities and growth, with particular reference to: electronic network development; the role of information in selected goods, services and labour markets; the characteristics of local information markets; and the distributional aspects of specific information and communication technology developments.

— The Centre for Communication and Information Studies (CCIS) has recently been formed at the Polytechnic of Central London. It brings together experienced researchers in various aspects of the information and communication technologies from Media Studies (including broadcasting, cable television, film and print). Law, Economics, Politics and Library Services (including the Library Technology Centre). The initial research programme will expand past work on markets and regulation in the audio/visual sector and the changing characteristics of library services. It will address evolving legal and policy issues concerning telecommunication regulation and the relations between information as a public good and tradable information.

— The Centre for Research into Innovation and the Culture of Technology (CRICT) at Brunel University will build on past research on the sociology of science and scientific knowledge; the public representation of science through the mass media; the economic and policy dimensions of product standard selection and the broadcast industries; and related work on the socio-cultural dimensions of technical design, and copyright law. CRICT will focus on the design, representation, marketing and use of ICT, and the relation of these factors to the professional, cultural, institutional, social and economic conditions within which technologies emerge and develop. The initial projects will reflect and build on the richness of their past research. The first is concerned with understanding the household as a site of ICT use. It will focus both on the way in which household use of ICT affects people's perceptions of ICT in other areas of their lives, and on the way in which ICT affects familial relations. The second project is concerned with the social determinants of software development, working from the hypothesis that the technical capacity, marketability and impact of software products reflect the assumptions of the designers and the contingencies of the design process. The project will be a comparative ethnographic study of two software development sites. The third project is a study of the demand for information and broadcasting products. It will involve the development of techniques for analysis and forecasting of demand. The specific products have yet to be determined, but possible examples include subscription television and an information service such as Prestel. The data, and the development of appropriate methods, will provide a basis for understanding and forecasting market relationships and for developing regulatory policies.

The Research Centre for Social Sciences at the University of Edinburgh will build on its expertise in the analysis of the ways in which new technologies emerge, how they relate to scientific advances and how they are affected by a range of political, organisational and cultural factors. Past research has focused on processes of technological development in research and development phases, the ways in which technologies are modified during implementation and diffusion and how feedback from these phases influences subsequent research. The first project will investigate 'organisational shaping in integrated automation'. It will examine the processes of social shaping during the implementation of advanced ICT systems in user organisations focusing on the ways in which organisational structures and the distribution of knowledge within firms constrain and pattern its development; and on the emergence of new models of work organisation. It is expected that a second project will be developed which will focus on the processes shaping the emergence of the shift from sequential to parallel computing. This would be a 'real time' longitudinal study of the emergence of a potentially radical technology. Choices made in the development of parallel computing will affect future developments in artificial intelligence in particular and 'fifth generation' computing generally.

— The Centre for Research on Organisation, Management and Technical Change (CROMTEC) at the University of Manchester Institute of Science and Technology will examine the implications of ICT for the changing structures, behaviour and performance of both public and private organisations. It will focus on: (a) the way in which organisations are choosing to use ICT to change the nature of internal information flows; (b) changes in the structure and organisation of the work of particular occupational groups; and (c) organisational interactions in ICT development and use. In order to maintain the integrity of the CROMTEC programme, the inital projects each address one of these areas. The first project will examine 'ICT and organisational change in the National Health Service'. Particular attention will be paid to inter-professional rivalries and organisational sub-cultures and their role in influencing the nature, use and consequences of particular information systems. The second is the first in a series of projects which will focus on the role of ICT in the financial services' industry. At this stage, the relative success of the design, development and use of ICT within different financial service companies in planning, management decision-making, job design, work flexibility and product innovation will be examined. The third focuses upon the competitive process and strategies adopted by firms attempting to secure and sustain a differentiated position in evolving ICT product markets — expert systems and value-added network services in the first instance.

An important component of the research programmes at each of the research centres will involve issues of mapping and measurement of some aspect of the information economy. Reliable and relevant sets of data constitute an indispensable foundation, both for sound scientific research into the structure, dynamics and social effects of the information economy, and for the development, implementation and monitoring of public policies and corporate strategies in the field of information. It is generally recognised that such data do not exist and that existing systems for the collection of statistics are becoming less and less adequate and relevant to the purposes for which they were initially created.

This research effort will build on past research done in the USA, France, Japan and other countries, as well as work done by international agencies such as the European Commission, OECD, UNESCO and other agencies. A major attempt will be made to develop new ways of conceptualising the field to be mapped, and to establish appropriate categorisations within which data can be collected and organised in a fashion that will provide an improved understanding of the information economy and its implications for other industries and public policies. This major research exercise is intended to prepare a substantially improved information base for more specific and directed research projects that require a sound information base on which to build.

Additional major projects in the initial phase of the research programmes at the centres include: (1) a comparison of policies and regulatory regimes in telecommunication, broadcasting and tradable information. Are they consistent with one another and achieving their objectives, or are they contradictory and counterproductive in some respects? How might they be adapted in the new information environment, with a much greater international dimension, to traditional national markets so as to achieve their objectives more effectively?; (2) an analysis of: (a) the use of information and communication technologies; and (b) trade in information services, in organisations in the northern region of the UK. This is a study of the diffusion of information technologies and services, with particular reference to geographical dispersion, and its implications for employment, economic growth and public services; (3) information technology and the military sector: implications for industrial competitiveness and economic growth. How does heavy spending on information technology, and information services, in the military sector stimulate and/or retard economic growth and competitive opportunity in the industrial sector? What are the implications?

15.7　PICT

An important aspect in establishing research networks is to provide opportunities for those research institutions that are not selected as information and communication technology research centres, and researchers working independently or in isolated locations, to benefit from the PICT. This benefit may come in several different forms. One form is the publication of the National Directory that provides access to information that otherwise would not be available to independent researchers. A second form is to take steps to involve independent researchers and other research institutions more directly in research networks. The practice of 'networking' is simply an efficient way to obtain information. Traditional forms of networking include conferences, seminars, journals, libraries, mail and telephone, all of which provide useful avenues for obtaining and/or exchanging information. For many kinds of research, electronic networking can facilitate research and communication significantly. An active networking role is included in the activities of the designated major research centres that will provide independent researchers with access to documentation, periodic seminars, collaborative research opportunities, and possibly other benefits. One task of the PICT during the second year is to assess the potential of electronic networks for enhancing the productivity of researchers and the effectiveness of research.

15.8 CONCLUSION

The ESRC PICT is an ambitious initiative. Yet there is both a need and an opportunity to provide a large quantity of useful policy research on a continuing basis, and to train a new cadre of professional researchers in the information and communications technologies field. The PICT is a response to this challenge.

By promoting research on long-term policy issues, the PICT may stimulate a modification of the traditional social science research culture by making information and communication exchange a more central part of research activity, and long-term policy research a field in which it becomes generally recognised that social scientists have much to contribute. For social science researchers, the implications of the information and communication technologies are no different than for the rest of society. They too must reassess their role in society and adapt in a manner that will permit them to contribute most effectively. They have much to contribute in preparing the ground for more informed public discussion on policy issues and more thoroughly considered policy decisions by government and industry. The PICT soon will be demonstrating this.

REFERENCES

[1] W. H. Melody, 'Implications of the information and communication technologies: the role of policy research', *Policy Studies*, **6**, p. 10, October (1985).

[2] See, Cabinet Office, 'Inmformation Technology Advisory Panel (ITAP), Making a Business of Information: A Survey of New Opportunities', September (1983) HMSO, London.

[3] For a review of telecommunication policy issues, see W. H. Melody, 'Telecommunication: policy directions for the technology and information services', *Oxford Surveys in Information Technology*, pp. 77–106, (1986) Oxford University Press, Oxford.

[4] M. Ferguson, ed., *New Communication Technologies and the Public Interest: Comparative Perspectives on Policy and Research*, (1986) Sage Publications, London.

[5] For a detailed exposition on this point, see Section 1 of W. H. Melody and R. E. Mansell, *Information and Communication Technologies: Social Science Research and Training: Volume 1 An Overview of Research*, (1986) ESRC, London.

16

New communications technology and regional inequality

Kevin Robins
Sunderland Polytechnic
Mark Hepworth
University of Newcastle upon Tyne, Newcastle upon Tyne, NE1 7RU, UK

16.1 INTRODUCTION

In business, political and media circles, the new information and communications technologies (ICTs) are widely and wishfully invoked as the panacea for economic decline and stagnation. It is commonly believed that we are on the threshold of a qualitatively new historical epoch: post-industrial society, the micro millenium, the silicon civilisation, the third wave, or whatever. Behind this dull and familiar discourse, we would argue, there is a compacted set of premises that equates technological progress with economic growth and human betterment [1]. Technology, progress and democracy are ideologically intertwined; it is an unquestioned and ossified article of faith that 'progress always result[s] from improved machines' and that 'progress [is] the handmaiden of democracy' [2]. This is the empty but powerful dogma that lurks at the heart of most discussions of the new technologies.

The essence of this pedestrian, futuristic orthodoxy is a belief that information can dissolve away social inequalities. In the new age, it is argued, information will be equally and plentifully available to all of us, and, in so far as information has become the source of wealth and plenty, so too will the riches of society flow freely and bountifully to each of its citizens. Information and communications networks are seen as supporting and encouraging the more equitable distribution of both wealth and power in the post-industrial society.

16.2 CENTRALISATION OR DECENTRALISATION?

Within the discourse of futurology, this idea is invariably related to the supposed shift from centralisation to decentralisation inherent in the new technologies [3]. The break up of concentrated and massified bureaucratic and industrial structures in favour of devolved and distributed organisational forms is seen as one of the progressive aspects of the new ICTs. (A key symbol of this technological promise is the 'electronic cottage' from which, irrespective of location, we shall be able to work, consume, play and communicate.) What is being suggested here is that the processes of decentralisation inherent in the electronic information grid can overcome and dissolve those geographical limitations of space, distance, location which, in the old industrial society, encouraged concentrations and inequalities in wealth and power. Now what is being addressed here is indeed an important issue: the spatial implications of technological change [4]. However, the difficulty and complexity of this issue is lost in a narrowly technicist emphasis which sees centralisation/decentralisation as simple alternatives, and which suppose that decentralised forms will unambiguously displace centralised organisational forms, just as post-industrial society supposedly succeeds industrial society. The spatial implications of ICTs, we would argue, are far more complicated and ambiguous than this. What is involved are questions of organisational restructuring; the ownership and control of space; the shifting significance of location; the superimposition of abstract (information) space on concrete (geographical) space; shifting boundaries of public and private space [5].

Although we dismiss the narrow technicism of this futurological perspective, we do indeed take very seriously the spatial transformations associated with the so-called information revolution. The new ICTs will profoundly restructure not just corporate and bureaucratic organisational forms, but also the broader UK space economy [6]. We cannot go into these matters in detail here. Our present discussion of the spatial implications of new technologies is more specific: to consider the obstacles which the 'actually existing space economy — and, particularly, the existing patterns of inequality at a regional level — present for the futurologists' dream of an 'information revolution'.

A perverse element of futurological discourse is the strange belief that the 'information society' is the consequence of an abrupt and absolute transition: like some giant butterfly, British Society will shed its chrysalis of industrial and manufacturing plant, and emerge, metamorphosed, in the post-industrial form of a pure information and service economy. It is remarkable that we should have to point out that no historical change is as clean-cut as Alvin Toffler and his like would have us believe. Britain inherits a solid and historical industrial infrastructure, and social practices and traditions almost as solid, and these cannot just be conjured away by the would-be alchemists of post-industrialism. For the myth of immaculate conception and virgin birth, we must substitute an account of the 'actually existing' context of economic and technological restructuring in the 1980s.

This context for the emerging 'information revolution' is a British (space) economy which is rapidly becoming an industrial wasteland, and which, moreover, is characterised by a particular geography of de-industrialisation. The reality is that 'regional inequality is widening, while almost the whole of UK economy is being transformed into a problem region of Europe at a pace that is currently concealed by the yield of the oil wells in the North Sea' [7].

In the 1980s, Britain is economically, socially and spatially a divided nation: 'There are increasingly two Britains separated by widening economic and social disparities. The northern and western regions have suffered disproportionately from de-industrialization with consequent adverse effects on output, incomes, the regional balance of payments and jobs' [8]. In this context, the North East fares particularly badly. According to the *Regional Development Programme 1986–1990* presented by the British Government to the European Economic Community in July 1986 (and still not published), there is no sign of economic recovery in the region's economy, and job dependence on stagnant traditional industries is higher than the national average; high technology industries are scarce; incomes, skill levels, educational achievements and health standards are below the national average; and the communications, water supply, land drainage, gas and water disposal infrastructure is badly in need of investment [9].

16.3 IMPLICATIONS OF TECHNOLOGY

As we have seen, there are those who confidently expect the information economy to be born miraculously, like some post-industrial pheonix, out of the ashes of British manufacturing industry. How, and indeed whether, this process of economic regeneration will occur is by no means clear. Let us emphasise that there are considerable (perhaps intractable) impediments. And these are all the more difficult in the case of a disadvantaged region like the North East. It is indeed the case, as the prophets of post-industrial deliverance argue, that the new ICTs can, *in principle*, overcome spatial and geographical inequalities. Information and communication networks do indeed provide the *technical* possibility of overcoming regional imbalances: information labour and other information transactions can, *in theory*, be freed from locational determinants and constraints. Such a perspective, however, betrays a narrowly technical and instrumental conception of social management. It does not confront those recalcitrant social, political and cultural forces that surround and inform technological development. It does not acknowledge the weight of historical conditions. In reality, society and economy, like the *Queen Elizabeth*, have their own ponderous inertia and are difficult to navigate.

As the information economy begins to take shape, we most certainly do not find that information work, information resources, and information wealth are becoming available to everyone, everywhere. The old locational and regional inequalities, brought about by the uneven development of capitalist industrial production, are not being superseded, but rather accentuated and transformed, by the exploitation of new ICTs. Social polarities are growing, and they are assuming new forms of spatial expression. What we are seeing is a growing regional dualism between the 'sunbelt zone' of the Greater South East — an area south of a line reaching from the Bristol Channel in the west to the Wash in the east — and most of the rest of the country.

Available data suggest that social and spatial inequalities are at the heart of the 'information revolution' (for fuller discussion, see [10]).

High-status information and service jobs are gravitating irresistibly to the south. Department of Employment figures show that around 94% of total jobs lost between 1979 and 1984 were in Scotland, the North, the Midlands, Wales and Northern Ireland, with the South suffering only a 6% loss. Moreover, more than two-thirds of service jobs created in this period have been in the South East. High-tech businesses are favouring sites in Southern Britain, with a concurrent swing away from urban centres towards rural areas' [11].

Other indices of the development of an information economy and infrastructure similarly reflect the subordinate and peripheral status of the North. The telephone and telecommunications system is clearly central to the establishment of an 'information society' in so far as it constitutes a foundation technology for any information network. It is, however, the case that, although the South East is already extensively wired-up, around one-third of the households in the North East are still without telephones. If there is some indication of a process of 'catching up', optimism must be offset against the likely consequences of future hikes in charges to domestic telephone users. The telephone will surely become a luxury item for a large part of the British population (cf. [12]) as ordinary phone calls become prohibitively expensive (let alone a new phone-based service like Prestel). The prospects for additional services like home shopping and tele-banking look all the more bleak. In the developing scenario of the information economy, the Greater South East will monopolise information labour and produce information services and commodities; the role of the North East within this new national division of labour can only be to provide a market for these same services and commodities. The preponderance of poorer and unemployed working-class people in the North East represents, however, a market barrier to the generalised availability of the new ICTs and the consumer services to which they grant access. The new technologies in the region may create the 'leisure' time for information consumption, but its historical condition of relative poverty, high unemployment and working class structure makes it a truncated market with a limited capacity for absorption and consumption.

While London consolidates its 'world city' status and joins Tokyo and New York as one of a handful of élite 'information cities' [13], the centrifugal force of the 'information revolution' throws the likes of Sunderland or Middlesbrough out into the economic wasteland. If, in principle, they become electronically 'closer', peripheral regions like the North East become, in reality, ever more adrift from the national and international centres of economic and political power. Real possibilities of self-determination and agency are further diminished as the region becomes subject to the new investment and location strategies of multinational capital. Given their powerful capabilities for radically altering the temporal and spatial parameters of economic life, what, for example, is likely to be the regional impact of new private computer networks for the intranational and international transfer of corporate and commercial information within and between firms? The North East economy, with its high level of dependence on externally-controlled branch offices and branch plants, is particularly vulnerable to changes in corporate geography which arise from computer network innovations in large multilocational firms [14].

While the new ICTs reinforce economic ties between the dominant 'information cities' of the British and world economies, what will happen to other urban and regional communities? Is the destiny of these communities to be the twilight world of an off-line economy devoid of jobs, market opportunities and decision-making power?

16.4   CURRENT INDICATORS

Some key facts indicate that corporate strategies are turning great tracts of Britain into backwaters of the information economy. For example, 75% of the newly-licensed value added networks are based in and around London, with other cities serving only as unmanned communications gateways to local business markets for on-line information services (coincidentally, an identical share of new jobs in Britain's information technology industries falls south of the great North–South divide). The fall out of the Big Bang, with its epicentre in the City of London's Golden Square Mile, adds up to only a few rads in the rest of the country, where the branch offices of the re-named International Stock Exchange, the high street banks, the building societies and other financial institutions compete for the foreign-bound personal savings of Britain's embryonic 'property-owing democracy'. At the same time, the arrival of Japanese and American financial conglomerates has tightened London's stranglehold of the British information economy — directly, by reinforcing the capital's status as a national and international head office centre, and, indirectly, by British Telecom's accommodation of the Big Bang through massive investments in the City's digital infrastructure of network technology. We can only conclude that, while London, the information city of Docklands, teleports and smart buildings, draws closer to New York and Tokyo, the capital's social and economic distance form the country's declining industrial areas in Northern England is increasing with every electronic bleep.

Is uneven development in the information economy inevitable? Those who subscribe to the orthodox economic law of comparative advantage, and believe in mythical perfect markets and fully-informed market actors, will doubtless claim that this 'iron law' operates in the information economy, much like it was supposed to have during the course of the Industrial Revolution. Within Britain, according to this law, indiviudal regions will (and should, if national output is to be maximised) specialise in producing and trading in those goods and services which best suit their respective endowment of resources. With the coming of the 'information age', however, three basic questions still need to be asked.

First, in so far as the geographical distribution of information resource endowments is not naturally determined (like, say, iron ore deposits) but historically determined by the country's experience of industrialisation, Britain's constituent regions are not equally well placed to take advantage of the commercial opportunities presented by the 'information revolution'. In the gold rush for information — as work material, services and commodities — some regions will have a head start. Can other regions, handicapped by their unique historical contributions to Britain's industrial development, really catch up by successfully 'making a business of information' — the Information Technology Advisory Panel's panacea for the nation's econmic ills? [15].

Second, if job prospects and industrial renewal throughout Britain now depend on access to information, is there really scope for balanced regional development in the information economy at all? If the Government were to introduce regional policies designed to achieve an equitable balance in the geography of economy opportunity, how effective would control over these adjustment policies be in the new open world of global factor and commodity markets? If the South continued to grow increasingly 'information rich', would the economic future of the North depend on the successful pursuit of a far riskier course of development — namely, re-industrialisation based on the application of information technology in manufacturing plants rather than the production of new information goods and services? Could, then, the North's computerised branch plants create sufficient numbers of the right types of job that are so badly needed to vanquish the region's chronic unemployment problem? The example of the new Nissan car factory, located in Sunderland, suggests that foreign-led re-industrialisation wll not provide a lasting and sufficient basis for sustained growth because of the Japanese plant's minimal production linkages with the rest of the regional economy. According to one source, the Policy Studies Institute, factory jobs threaten to disappear at a faster rate in the North East because microelectronics technologies are primarily used in the form of process innovations (shown by the PSI to reduce employment levels) rather than product innovations (identified as job-generators) in the region's manufacturing firms.

Third, accepting orthodox economists' contingency assumption that regional structural adjustment takes time, how long will different parts of Britain need to wait before the benefits of the 'information revolution' reach them? Can governments and industrialists believe that the trickle-down effects of the 'revolution' (redistribution with growth) will outweigh its polarisation effects (concentration with growth), making regional underdevelopment but a transitory problem or by-product to be addressed by policy makers at a later date? Unfortunately, guaranteed futures have become part and parcel of the political rhetoric of the 'information revolution', and, like most marketplaces for perishable information, the guarantees offered by 'experts' come with escape clauses — the questionable 'all things being equal' assumptions.

16.5 REGIONAL IMPACT

The impact of the new ICTs on the regional geography of Britain is proving to be profound and divisive. To a large extent the necessary policy implications will depend on one's understanding of the nature of technological and economic development. There are those, for example, who see divisions and inequalities as only temporary problems associated with the early stages of developing any new technology. For Benjamin Compaine, technological innovation starts 'with a small vanguard of adopters who tend to be better off economically than the population at large . . . the market created by this vanguard often starts a process which leads to greater interest, higher volume, thus lower cost, reduced skill levels needed, and ultimately mass utilization' [16]. In his view, technological development is associated with declining costs and a wealthier work force, and, consequently with a 'lessen[ing of] the difference in life style between the poorer and richer in society' [16]. This faith in the (ultimately) egalitarian bias of new technologies draws, we would suggest, on that accepting equation of technology, progress and democracy, which has become received and commonsense wisdom, and which pervades most discussion of the 'information revolution'.

Technology is seen as a neutral, but generally benevolent, force. Inequalities remain a temporary and subsidiary problem that can be resolved through further technological progress and development. Far from being problematic, technological growth is seen as the panacea for all social and economic inequalities, and this, in so far as it is, the major contributor to economic productivity and wealth creation. In the context of our discussion of ICTs and the regional problem, this conception of technology and technological development sees inequalities as nothing but 'hitches' in the process of technological diffusion. The policy implication is to 'iron out' any problems and to facilitate the inherent and inevitable tendency of technological development to gravitate towards economic growth and equality.

This is the received wisdom. In reality, however, divisions and inequalities persist, and the objective of equality and democratic fulfilment always remains a remote promise receding into an elusive future. The fundamental problem with this approach is that technology is seen as a 'fix', a neutral instrument for social amelioration, and, as such, it retains a pre-social innocence and purity. What it cannot, and does not, address is the fact that technologies are implicated in the very ills they are supposed to anneal. Technologies are inherently and intrinsically social: they mediate and express the (unequal) relations of political and economic power that characterise our society. One manifestation of the unequal social development that has been brought about by technological growth under conditions of capitalist accumulation has been the (spatial) process of uneven development (on both a national and an international scale).

An important, but neglected, emphasis on the inherent 'bias' of technology, and particularly its spatial bias, is contained in the work of the Canadian economist, Harold Adams Innis. The special relevance of his work lies in his central concern with the social implications of communications technologies. What Innis recognised was that 'new competing communications media alter the forms of social organiza- tion, create new patterns of association, develop new forms of knowledge and often shift the centres of power' [17]. Only if we consider the inherent bias of all technologies, only if we recognise that technologies mediate and express the prevailing social and power relations shall we be in a position to confront 'biases' that modern information and communication technologies are likely to impose on the so- called "information societies" of the future' [18].

16.6 CONCLUDING REMARKS

An adequate conceptualisation of the new ICTs (and of technology generally) is vital to the policy process that now confronts us [19]. The social implications of these new technologies are far more complex and difficult than is suggested in the discourses of industrialists, politicians and the pundits of post-industrialism. Intervention in policy issues requires more than capitulation to technological inevitability, oiling the wheels of 'progress', and then anticipating that economic productivity and growth will salve (or defer?) actue social problems. The development of ICTs is occurring in a divided and de-industrialising Britain. The indications are that, far from electroni- cally dissolving the legacy of the Industrial Revolution, the new technologies are in fact reinforcing them. One fundamental aspect of this process, we have argued, is the intensification and recomposition of regional (and urban) inequalities and polarities. Only in the most trivial technical sense do the new technologies overcome the tyranny of geography. The information economy in Britain is emerging as 'a core–periphery pattern of spatial economic differentiation' [20] in which regions like the North East drift into remote orbit, whilst the Greater South East, as the headquarters of internationally–orientated capital and finance, becomes the control centre of the information economy. The North is suffering acutely from the problems of manufacturing decline and de-industrialisation, but it is far from clear that there will be an electronic salvation. What seems to be the case is that the ways in which the emerging information economy is 'being organised over space mean that geographical inequality is actually inherent is the spatial structure itself . . . What need[s] to be said about the new form and nature of geographical inequality in Britain is that it is *integral not to decline but to growth*' [21]. The problems of the North East are not simply unfortunate residues of industrial and manufacturing decline. This is, in part, the problem, yes. But of far more concern is that the very spatial dynamic of the information economy itself is reinforcing that process of uneven development which is at the heart of the regional problem. Spatial and social polarisation, it seems, is an expression, not of 'dysfunctions' in the information economy, but of its very effectiveness and success. The problems this poses for social and regional policy are ones we still have to confront in all their seriousness and complexity.

REFERENCES

[1] K. Robins & F. Webster, 'Information technology: futurism, corporations and the state', in R. Miliband and J. Saville (eds), *The Socialist Register,* (1981) Merlin Press, London.

[2] E. S. Ferguson, 'The American-ness of American technology', *Technology and Culture,* **20,** 3–24 (1979).

[3] F. Webster & K. Robins, *Information Technology: a Luddite Analysis,* (1986) Norwood, NJ, Ablex.

[4] M. Hepworth, 'Information technology as spatial systems', *Progress in Human Geography,* **11,** No. 2, forthcoming.

[5] M. Hepworth & K. Robins, 'Home interactive telematics and the urbanisation process', Paper presented to the *International Conference on the Social Implications of Home Interactive Telematics IFIP/NGI,* Amsterdam, 24–27 June, (1987).

[6] J. B. Goddard, 'Technology forecasting in a spatial context', *Futures,* **12,** No. 2, April, 90–105 (1980).

[7] M. Dunford & D. Perrons, 'The restructuring of the post-war British space economy', in R. Martin and B. Rowthorn (eds), *The Geography of De-Industrialisation,* (1986) Macmillan, London.

[8] J. Rhodes, 'Regional dimensions of industrial decline', in R. Martin and B. Rowthorn (eds), *The Geography of De-Industrialisation,* (1986) Macmillan, London.

[9] A. Moreton, 'Unemployment blackspot with poor outlook', *Financial Times,* 3 November (1986).

[10] M. Hepworth & K. Robins, 'Whose information society? A view from the periphery', *Media, Culture and Society,* forthcoming.

[11] P. Marsh, 'The lure of the silicon glen', *Financial Times,* 12 November (1986).

[12] R. Pike & V. Mosco, 'Canadian consumers and telephone pricing: from luxury to necessity and back again', *Telecommunications Policy,* **10,** No. 1, March, 17–32 (1986).

[13] M. Moss, 'Telecommunications policy and world urban development', Paper presented to the *Annual Meeting of the International Institute of Communications,* Edinburgh, September (1986).

[14] M. Hepworth, 'The geography of technological change in the information economy', *Regional Studies,* **20,** No. 5, 407–424 (1987).

[15] Information Technology Advisory Panel, *Making a Business of Information,* (1983) HMSO, London.

[16] B. M. Compaine, 'Information gaps: myth or reality?, *Telecommunications Policy,* **10,** No. 1, March, 5–12 (1986).

[17] W. H. Melody, 'Introduction', in W. H. Melody, L. Salter & P. Heyer (eds), *Culture, Communication, and Dependency: the Tradition of H. A. Innis,* (1981) Norwood, NJ, Ablex.

[18] W. H. Melody, 'Some characteristics of knowledge in the information society', Canada House Lecture, (1986).

[19] R. Negrine, 'The new information technologies: is there an "alternative strategy"?', *Capital and Class,* No. 31, Spring, pp. 59–78 (1987).

[20] P. J. Damesick, 'Recent debates and development in British regional policy', *Planning Outlook,* **28,** No. 1, 3–7 (1985).

[21] D. Massey, *Spatial Divisions of Labour,* (1984) Macmillan, London.

17

Toward a communication–information model: a theoretical perspective for the design of computer-mediated communication systems

K. L. Hacker and L. Monge,
Department of Humanities, Michigan Technological University, Houghton, Michigan 49931, USA

17.1   INTRODUCTION

We no longer need to hear about the wedding of communications and computing. The couple have not only been together for several years, they have already had children and grandchildren. From the parents of telecommunications and data processing have come the children of voice–data networks, teleconferencing, electronic mail, and other modes of computer-mediated communication. These results have produced what cannot properly be classified as either information systems or communication systems. Rather, they combine both functions, and are therefore communication–information systems.

This chapter argues that the design and implementation of communication–information systems is going on atheoretically in terms of communication. That is, communication–information systems are based on models of information transmission alone, without sufficient regard for (a) the complexities of human communication and (b) the interrelationships of communication and information processes. The unfortunate consequence of this situation is a fundamental blockage of progress in the evolution of human communication. That evolution should concern not only communication scientists and others who study human behaviour and society, but as importantly, if not more importantly, the designers of the technology which transforms the evolutionary processes. Gone are the days when the technical person could ignore the effects of research and development, or expect others to resolve them.

As some computer scientists have pointed out, computers are transforming our modern world. This combined with the critical nature of communication processes not only to the production of information but to the adaptation of entire cultures, we are in real need of theories to guide the planning of new communication–information systems [1].

We have too many models that do not account for the substance of what they are modelling; without that substance, we are missing many critical components of what we are working with.

Although it can be argued that the design of communication systems goes hand-in-hand with the discovery and research processes of human communication research, it must be recognised that the technologies have been developed far more rapidly than they can be completely understood in terms of effects. With television, for example, after hundreds of studies and decades of experiments, the relationship between televised violence and violent acts in society is only beginning to reach a point of explanation (stimulating effects theory).

Because communication technologies create mediation systems not only of data transmission but also of meaning construction, relationship formation, and cultural reproductions, the technologies have powers upon human organisation and adaptation that cannot be left to the whims of either technological imperatives or movements of supposedly benign market forces.

Today, there is a great deal of confusion about communication. Ask ten people what communication is and you will get ten differing answers. Ask a computer scientist what it is and you will learn about gateways, networks, and data streams. Ask a communications engineer, and you will learn about transmission, speech paths, and signal-to-noise ratios. A communication scientist may parrot some of these terms, but will be likely to add terms such as uncertainty reduction, organisation, and transaction. All three groups of scientists confound two basic things: (a) types of communication, and (b) communication and information. The types of communication that are not distinguished and clarified are (1) non-mediated human communication, (2) human–machine (computer) communication, and (3) human––machine–human communication. We are concerned here with the last-named.

There is also little clarity regarding the singular and interactive concepts of communication and information. They are often treated synonymously. In this chapter, we present a model of communication–information which separates the critical properties of both and shows how they interrelate. This model is based on the empirical study of human communication. It is neither philosophy nor extensions of untested maxims. Although not a theory, it is a starting point toward improved models and new theories which may link observations of computer-mediated communication (CMC) systems to the development of the models.

17.2   THE TECHNOLOGICAL IMPERATIVE

Communication scientists have generally followed the development of communication technologies, attempting to explain their effects and suggest new uses and directions for technical developments. Generally speaking, they have been left out of the design process. Rogers [2] says that this is changing. The change began with the technology of videotext. According to Rogers, Elton and Carey tested a teletext service for a year and concluded that the designers of the technology did not understand the existing information habit of the public. Their study led to redesign and to suggestions for how people might adopt the technology.

The results of atheoretical communication–information systems design are not academic. Generally, CMC systems are not used as much or as efficiently as they would be if they were more effective in design. Managers and users are often confused about CMC objectives, control, and planning. Managers, for example, do not know who in an organisation should have responsibility for the networks [3]. Most management decisions about CMC technology are based on economic hypotheses [4,5]. Most organisational planning of communication technology has been *ad hoc*, accomplished with little comprehensive planning [6]. 'Office Automation' and 'Office-of-the-Future' are rhetorical calls that were once death-knells to the paper-drowned office. Today, offices are still filled with paper. People have anxiety about not being computer literate—a condition they have perceived as second only to venereal disease. Home computers are bought with elaborate software packages and used for playing simple-minded games. Videophones roll off the drawing boards into communications museums. Workstation dreams turn into chapter 11 nightmares. Schools buy computers as quickly as possible despite the lack of evidence that computer programs enhance either intellectual functioning, problem-solving, or motivation [7]. Businesses rush into computer and CMC systems to fulfil dreams of heightened productivity. Some of what they were promised turns out in fact to be misleading. As technology followed its own lead, and everyone else followed technology, computers became less user-friendly than people became computer-friendly.

When led by the technological imperative, the design of CMC systems results in systems which inadequately facilitate human communication. In turn, because of the interdependence of information and communication, information processes are also insufficiently aided.

17.3   GOOD IDEAS GONE BAD

The current vision of communication–information systems is summarised in the term Integrated Services Digital Network (ISDN). This vision details the various ways that telex, telephone, music, video, and even holograms can be shared by global communicators. The benefits are legion, including such services as medical patient monitoring done with a blend of telemetry and data processing. Gantz [8] throws some rain on the parade, however, by pointing out some of the problems with ISDN. First, there is little certainty regarding the use rates of lines. Second, if voice–data bandwidth ratio projections continue to be 10:1, there is doubtful need for the massive digitalisation that is being heralded. Third, the communication uses of communication technology users are not sufficiently understood. Remember the Picturephone of 1964? Videotext? Home banking, personal stock quotations, home electronic mail, and other Information Age services were appealing on trial bases, but easy to reject in real money terms. According to Gantz:

"Pioneered by government-owned PTTs in England and France, the initial trials—where equipment and programming were free—were successful enough to get the agencies to commit to wholesale fallout. Alas, in the real world, where subscribers must partially defray costs of termination units and supplied services, nobody seemed interested in marching through the turnstile of the Information Age. At least, not when the ride wasn't free."

Satellite Business Systems was sure that businesses were ready for large-scale high-speed data transmission capabilities, including teleconferencing. Instead, users wanted cheaper long-distance phone calls, and what SBS became was another MCI or SPRINT. Gantz argues that other failures appear to be cellular phones, air-phones, dial-up electronic mail, and multi-tenant services. Today, less than 50% of American central offices terminate in digital switches. Personal computer companies have found real problems in the 'electronic cottage' market. Mitel and other PBX manufacturers are wondering why users do not use the three hundred phone features they offer. Mitel wonders why the SX-2000 did not fly.

Communications systems are being rapidly developed and rapidly reinvented as applications are found wanting in any real improvement beyond greater connectivity. According to Ralston and Reilly [9], most MIS failures are due to: (a) poor identification of user needs, (b) over-rigid applications, (c) domination of technical over other factors during design, and (d) lack of attention to human and social factors. We argue that although these are true, and although it is also true that many new communication technologies appear to be failing, the real issue goes to the origins of all this difficulty; that is, communication is being technologised without a fundamental understanding of the central operating principles of communication.

17.4 COMPUTER SCIENCE AND HUMAN CONDITION

The primary concern of computer science has been defined by Ralston and Reilly as a focus on 'information processes, with the information structures and procedures that enter into representation of such processes, and with their implementation in information processing systems'. Perceiving a need to guide the growth of computers, bring order to the uses of computers, and steer new designs and applications, computer scientists focus on structures, operations, principles of design and programming, and methods of use in different classes of information processing. With the development of communication–information systems, computer scientists have become more involved with communication. Communication is basically conceived of as the processing or handling of information.

Both computer science and communication science have roots in information theory. What is problematic is the bifurcation of communication theory that has occurred—one being the development of human communication theory, the other the development of machine communication theory. Both of these sciences treat human communication in descrepant ways. Communication science studies dynamics of relationships, meaning formation, action development, rules, and interpretation. Computer science looks at the signals, circuit switches, message paths, terminal access to mainframes, and load sharing of data processing. Both have missed the connections that occur when people communicate through the machines

(CMC). Fortunately, this problem is being approached more and more through research done with the resources of both disciplines. For example, researchers from both areas meet and share ideas each year in the Human Communication Technology group of the International Communication Association.

Hacker and Monge [10] have made a content analysis of IBM and AT&T design literature to determine how designers conceptualise communication. The results of this study indicate that the two most prevalent concepts about communication are (ranked): (1) connectivity, and (2) sending–receiving. The lowest two are interpersonal interactions and personal relationships. It appears that CMC designers do not distinguish data transmission from human communication, nor machine communication from human communication. It is also evident that communication is treated more as mechanical transportation than as an active human process of interaction with self, others, and the environment.

It is interesting to note that some computer scientists see communication as a process subsumed in larger processes of information-handling. Ralston and Reilly, for example, claim that 'the main concern of information science is with processes of communication, storage, management, and utilization in large database systems. Thus, the domain of information science is included in the broader domain of computer science'. What appears to be operating is a basic confounding of transmission with communication.

17.5   THE POLITICAL EFFECTS OF INFORMATION-BASED COMMUNICATION MODELS

A computer scientist once told a 1985 gathering of communication technology theorists that he could design any kind of political system imaginable; it could be a democracy or it could be a dictatorship. What is significant about his comments is not only that such political design is possible through the use of communication technology, but that the source of new political structures has three possible origins: (1) technical imperatives which naturally lead to technocratic elitism, (2) *laissez-faire* neglect with the understanding that someone else will think out the problem, and (3) a new commitment to democracy with a sensitivity to the impacts on democracy by the new technologies. The directions today appear to point to greater centralisation (albeit accompanied by rhetoric of decentralisation), greater systems of behaviour and cognitive controls, and less personal privacy. It is interesting to note that Honeywell and others promote the same linguistic connection between 'Communication' and 'Control' that the Pentagon does.

Ellul [11] notes how media have been able to force habits of self-defeat on individuals through the socialisation of 'technique'. He argues that as people become accustomed to listening and talking machines, they become indulged in technical procedures at the price of damaging their connections with themselves and other people. He says that consciousness fuses with an omnipresent technical diversion.

As did Plato in his formulation of the utopian Republic, modern communications modelling assumes that better designs will facilitate better social functioning. Today's knowledge-workers and programmers thus become akin to Plato's philosopher-kings. They are the ones who have the light and should therefore hold power.

Boguslaw [12] notes that the 'clock of the contemporary world' is dedicated to the idols of physical efficiency and that power is redistributed through the development of system control mechanisms:

> "Our own utopian rennaissance receives its impetus from a desire to extend the mastery of man over nature. Its greatest vigor stems from a dissatisfaction with the limitations of man's existing control over his physical environment. Its greatest threat consists precisely in its potential as a means for extending the control of man over man."

Tehranian notes some of the fool's gold in the glitter of the Information Age. According to Tehranian [13], the age of information has produced isalnds of information abundance, information resource inequity, and widening gaps between groups. In addition, he argues that the primary beneficiaries of the new communication technologies have been the military and transnational corporations. Instead of creating cultural diversity and creativity, the new communication technologies have fostered greater bureaucratisation, more elites with greater powers of domination, and ideological predominance of programming efficiency above anything else. Communication has become more hierarchical, damaging to personal privacy, and greatly routinised.

The information age has been what it is because of its epistemological roots in information theory. That theory has promoted communication models which have created passive views of audiences, news that is objective, teachers who can toss knowledge out to hungry students, and politicians who can create systems of disinformation because of their influence over mass communication channels. Brokaw [14] notes an ongoing political effect in universities where students are forced to use computer programs that contain personal to guide the users through steps of writing. Brokaw notes that students in these conditions become adept more at following instructions than at critical higher-order reasoning processes.

Computer scientist Michael Conrad [15] presents a useful theorem for communication system design. According to his 'trade-off principle', any system loses evolvability to the extent that it becomes programmable. Structural programmability of a system is directly related to how much a system can formally be mapped in terms of components. This programmability entails total control through formal rules and procedures. A profound political struggle emerges in the structural progammability of CMC. To the extent that communication must follow technique, protocol, procedures, rules, and hierarchies, it becomes routinised, controlled, and limited. Worse yet, it becomes disempowering to the degree that it slows down the evolution of human communication. The evolution of human communication has been a progression from basic data-gathering interactions to higher levels of interpretation and meaning [16].

17.6 INFORMATION THEORY

Shannon [17] set the foundation for communications engineering with his theory of 'communication'. Information theory established a mathematical approach to communication described by communication engineering theorist Johannes Peters [18] in the following manner:

"A transmitter disposes of a set of symbols (a); a receiver is sensitive against a set of symbols (b). There is communication if the symbols (b) are statistically dependent on (a). This property to be statistically dependent is represented by a channel."

Probability theory is used to formulate channel capacities. According to Peters, information theory is not empircally based but is 'pure mathematics since it does not originate from empirical facts but from abstract definitions alone'. Although many communication theorists (human communication) recognised the problems with the electronic metaphors, many used the language of Shannon's model to describe what they thought was human communication—the sending and receiving of messages. Others have noted the erroneous application of information theory to human communication. Casagrande and Casagrande [19], for example, say that 'Shannon and Weaver were not talking about human communication; but early theorists saw the application possibilities. By today's standards, the model is inaccurate'. Shannon later said that he was talking about mechanical systems.

Littlejohn [20] notes four basic problems with applying information theory to human communication. First, the original information theories dealt with more signal transmission than with communication. Second, many of the codes that humans use in communication cannot be broken down into bits of information. Third, meaning tends to be downplayed in information theory. Without meaning, messages may have little use or value. Fourth, information theory avoids contextual factors of communication.

Like communication science, information science has a subject of inquiry that is critical to the design of CMC systems. Neither field has enough current theory to offer the engineers. Part of the problem is that two interrelated, interdependent processes are studied separately while systems design needs the findings of both areas.

Information theory has been very useful for electrical engineering, traffic engineering, and certain signal processing. However, the processes of human communication cannot be modelled, except in the most primitive ways by information models.

17.7 COMMUNICATION–INFORMATION

Information theorists have ebbed and flowed when it comes to the meaning aspects of communication. Travers [21], for example, says that 'information theory does not necessarily have anything to do with the communication of either meaning or knowledge'. Weaver [22], however, who was Shannon's co-theorist, suggested that meaning along with technical and effectiveness levels of information was of concern to a mathematical theory of communication. Despite Weaver, designers of CMC systems have chosen to ignore meaning. Information is treated in terms of uncertainty or entropy as realised in binary digits. Communication is treated as the transmission of information.

The mechanistic notions of Shannon's information–communication model are incosistent with empiral studies of human communication. Consequently, in the field of communication, the information models, whether at micro or macro levels, are being increasingly challenged by contextualist models which stress adaptation, changing forms of mediation rather than static channels, and the social construction of meanings.

Georgoudi and Rosnow [23] say that the central focus of the contextualist theories of communication is on the unity, plurality, and ecological dependency of human action. Rather than describing communication as directional, it is treated as active, reciprocal, and interactive. Communication is a relational process, not one of connection and transmission. The further one strips messages from their context, the more one makes the communication meaningless.

According to Georgoudi and Rosnow, communication is dialogic and a process in which the 'content or meaning of which is interpreted and reconstructed within a context of shared meaning'. Hacker [16], from empirical studies of human communication, has observed communication sub-processes of coorientation, self-expression and testing, coactive and interactive interpretation, structuring of significance, and emerging summary structures of discourse.

Communication scientists who study CMC systems have observed numerous characteristics of CMC. Some of their major finding are summarised below:

1. The development of social meaning, which requires non-mediated interaction, is necessary for coordinating information, complex tasks, and motivation [24].
2. Predicting uses of CMC systems in organisations requires a multivariate approach to context, attitudes toward technology, roles, and system infrastructures [25].
3. Although specific forms of CMC systems, such as electronic mail, CBBS, and computer conferencing, differ in attributes and features, they are similar in how they attempt to facilitate human communication [25].
4. Designers of CMC hardware and software do not cite current human communication theory in their literature [26].
5. Major design concepts include channels, messages, networks, nodes, operations, protocols, connectivity, and permeability [27].
6. CMC is inappropriate for some interpersonal tasks [28,29].
7. Emotional dimensions of communication do exist in CMC [30].
8. Some CMC modes such as electronic mail are not appropriate for private communication [29].
9. CMC systems may help one type of communication while hindering others. For example, upward message flow may be helped while lateral message flow are interpersonal relations are hurt [31].
10. CMC increases connectivity [28].
11. Communication skills and rules are related to satisfaction with communication technologies such as teleconferencing [32].
12. Social concerns such as face-saving remain in CMC interactions [29].
13. Perceived needs and current communication requirements influence a user's willingness to experiment with new technology [33].
14. Some interactive media, such as two-way cables, can suppress idea diversity [34].

These findings should indicate the following to us. CMC is not a replacement for non-mediated communication. It is there to facilitate the essential processes of human communication while preventing some of the problems that might occur in non-mediated communication (such as too many things to talk about at once). CMC systems are designed to aid communication, not just information [35]. It should also be apparent that connectivity and permeability are not enough for the design of these systems. Emotions, face-saving, relationships, and other non-data variables must be worked with. Information is what is worked with in communication processes, much as language is. In fact, information can be seen as the application of linguistic dynamics to data.

17.8   ALTERNATIVE MODELS

Discussing the 'ultimate communication system', communications engineers Pierce and Posner [36] state that 'advances in communication either begin with new, higher-performance hardware that makes new systems possible, or with concepts for new functions that can be performed with already available hardware'. Another argument has developed in this chapter. That argument is that the best communications systems will look deeply at human communication and all its complexities, and will then derive appropriate, fitting, and workable models for technical facilitation of these processes. The optimal (never ultimate) systems will be those that promote the natural evolution of the intelligence and humaneness of our communication abilities.

Meek [37] claims that 'It is the marriage of information processing with telecommunications which will restructure the world—not either on its own'. This chapter argues that the most productive and the most politically progressive restructuring will follow the development of valid communication–information models and theories.

Information-theory models suggest networks, nodes, and channels. This creates physical architectures and topologies necessary for data transmission and data interaction, both of which are phases of communication. Meaning models of communication stress interactivity in messaging (coding/decoding/coding as opposed to coding, then decoding), variations of message production processes, adaptation in choosing communication forms or media (mediation forms vs. transmission channels), and audience situations which are both active and coactive with the messaging process. As stated earlier, these two theories need to be brought together, so that the basic channel models of information theory can be integrated into the processual communication models of communication theory. When this is done, information takes its rightful position as a *component* of communication, and we thereby give each concept greater clarity and utility. In doing this, we can draw upon the Communication–Information Model (see below) to provide a starting point for moving beyond oversimplistic notions of communication.

17.9   COMMUNICATION–INFORMATION MODEL

This model is based on the empirical study of human communication, and provides the following principles that may become useful in the design of computer-mediated communication systems.

1. Messages are exchanged, transacted, and reconstructed through various mediation forms—letters, phone calls, electronic mail, meetings, etc., etc.
2. Communication is not something sent through a pipeline or transported across anything—it is the process by which data are transmitted and then processed in many complex ways.
3. Data (not information, not communication) are transported through channels in sending–receiving activities.
4. Each communicator has databases and while communicating shares or exchanges data while drawing upon the databases. Note that databases existed before computers.
5. Data sets interact in the process of human communication. The human communicator takes data and *makes sense* out of them. This sorting out, sense-making activity makes data coherent, organised, and useful. We call this process *interpretation*. Research indicates that interpretation is both coactive and interactive.
6. Through the process of interpretation (coactive and interactive), we develop *meaning*. Meaning is not a philosophical concept here; it is empirically observable as the contextualisation of data, the setting of relations between data elements, the structuring of boundary conditions, and the emergence of significance for salient conclusions.
7. As meaning is constructed through cycles of interpretation, four primary results emerge: uncertainty reduction, communicator relationships, information, and cybernetic adjustment.
8. *Uncertainty reduction* is the interaction of cognition with communication as meaningful information is related to ongoing plans of actions or activities.
9. *Communicator relationships* are the patterned interactions across time which bring together communication styles, habits, etc., into predictable and useful ways. These relationships oversee mediation channels and define their existence.
10. *Information* is interpreted as data made meaningful.
11. *Cybernetic adjustment* is action adjustment based upon the use of the communication process to produce information which can be fed forward as data and fed back as information.

This model, illustrated in Fig. 1, of communication–information is by no means complete or immune from debate. In fact, a good model probably stimulates some debate. The strength of this model lies in its origin . . . the empirical study of human communication. Its utility is its challenge to information models of communication.

17.10 CONCLUSIONS

Williams [38] says that 'if we are evolving toward a preindustrial society, an information one, or a communication age, it is the technological advances in computing and communication that are seen as driving forces'. He may be correct. If he is, there is much work to do in making the communications technology fit the communication–information processes that are most desirable for our cultures, societies, and organisations. This will undoubtedly require alternative models of

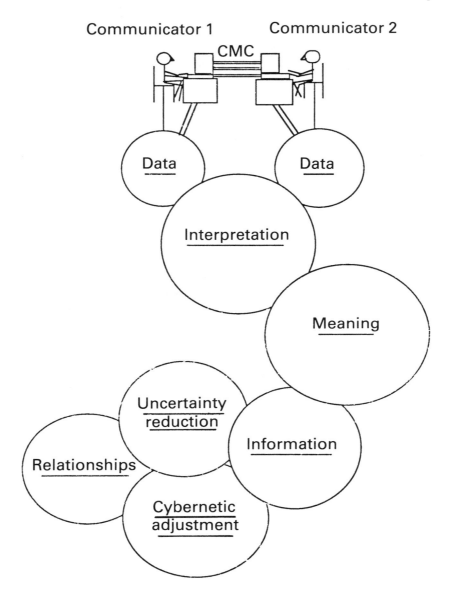

Fig. 17.1 — Communication–information model for CMC systems.

communication–information which fit empirical circumstances of communication and information evolvability.

Until fundamental issues of inadequate theory, political power struggles, premature modelling elegance, and control vs. liberation are dealt with, the technological imperative of the CMC system will continue, and with it the emergence of electronic Leviathans operated by programmer-kings.

Taking into account what has been learned in the empirical study of human communication and the suggested communication–information modelling, certain key conclusions can be drawn. First, CMC systems have followed models based on data transmission models of communication. These designs are inadequate not only for their limited application to how people actually communicate, but also due to their damaging political effects. Second, information theory and communication theory are compatible when it is realised that communication is the overseeing process which shapes and develops information. Third, models of human communication should be integrated with models of CMC. Fourth, the age of information should not be viewed as anything more than rapid advances in data processing and information. It should instead be seen as the stepping stone toward a more enlightened period of human evolution one might call the Age of Communication.

In a communication age, one should expect much more interactivity and self-determination of mediation controls. Rather than deskilling workers and managers, we might empower them with greater tools of interaction, interpretation, coaction, and meaning production.

The outcome clearly depends not on our stars but on our choices.
—Tehranian, 1985, p. 17

REFERENCES

[1] J. Weisenbaum, 'Nova', *PBS*, 14 April, (1987).

[2] E. Rogers, *Communication Technology*, (1986) Free Press, New York.

[3] A. J. Mallia, 'Who should manage office technology?', *Management Technology*, pp. 44–47, August (1983).

[4] L. Gould, 'Extending executive reach via communications', *Telecommunications*, August (1981).

[5] J. E. Bowes, 'Mind vs. matter—mass utilization of information technology', in B. Dervin (ed.), *Progress in Communication Sciences*, pp. 52–72, (1980).

[6] J. J. Connell, 'Is there an office of the future?', *Management Technology*, May (1983).

[7] D. Noble, 'The underside of computer literacy', in M. Gurevitch and M. R. Levy (eds), *Mass Communication Yearbook*, Vol. 5, pp. 585–612, (1985) Sage, Beverly Hills.

[8] J. Gantz, 'ISDN: How real? How soon?', *Telecommunications Products and Technology*, pp. 33–54, January (1986).

[9] A. Ralston & E. D. Reilly, *Encyclopedia of Computer Science and Engineering*, (1983) van Nostrand-Reinhold, New York.

[10] K. Hacker & L. Monge, 'Assumptions about human communication in the design of computer-mediated communication (cmc) systems: a content analysis of communication technology system engineering journals', working paper, (1987) Michigan Technological University, Houghton, MI 49931.

[11] J. Ellul, *The Technological Society* (1964) Vintage, New York.

[12] R. Boguslaw, *The New Utopians*, (1965) Prentice-Hall, Englewood Cliffs, NJ.

[13] M. Tehranian, 'Communication and empowerment: the dialectics of technology and mythology', Paper presented to the *International Communication Association*, May (1983) Dallas.

[14] M. Brokaw, 'Human–computer interaction and writing processes', Master's thesis (1987) (unpublished), Michigan Technological University.

[15] M. Conrad, 'On design principles for a molecular computer', *Communications of the ACM*, **28**, 464–480 (1985).

[16] K. Hacker, 'The need for meaning-centered theories of human communication: the paradigm shift from information theory to contextualism', Working paper, (1987) Michigan Technological University, Houghton, MI 49931.

[17] C. E. Shannon, 'A mathematical theory of communication', *Bell System Tech. J.*, 379–423 (1948).

[18] J. Peters, 'Entropy and information: conformities and controversies', in L. Kubat and J. Zeman (eds), *Entropy and Information*, (1975) Elsevier, Amsterdam.

[19] D. O. Casagrande & R. D. Casagrande, *Oral Communication*, (1986) Wadsworth, Belmont.

[20] S. W. Littlejohn, *Theories of Human Communication*, (1983) Wadsworth, Belmont.

[21] R. M. W. Travers, *Man's Information System*, (1972) Chandler, Scranton, PA.

[22] W. Weaver, 'The mathematics of communication', *Scientific American*, July (1949).

[23] M. Georgoudi & R. L. Rosnow, 'The emergence of contextualism', *J. Communication*, **35**, 76–88 (1985).

[24] A. Picot, 'Office technology: a report on attitudes, and channel selection from field studies in Germany', *Communication Yearbook* 6, (1982) Sage, Beverly Hills.

[25] C. W. Steinfield, 'Computer-mediated communication systems', *Annual Review of Information Science*, **21**, 167–202 (1986).

[26] K. Hirsch, K. Hacker & C. Carmichael, 'Design problems in computer-mediated communication-systems: implications for the field of communication', Paper presented to the *International Communication Association*, May (1985) Honolulu.

[27] R. H. Miller & J. F. Vallee, 'Towards a formal representation of EMS', *Telecommunications Policy*, **2**, 79–95 (1980).

[28] R. E. Rice & D. Case, 'Electronic message systems in the university: a description of use and utility', *J. Communication*, **33**, 131–152 (1983).

[29] G. Heimstra, 'You say you want a revolution? Information technology in organizations', in M. Burgoon, (ed.), *Communication Yearbook* 7, (1983) Sage, Beverly Hills.

[30] A. Phillips, 'Computer-conferencing: success or failure?', *Systems, Objectives, Solutions*, **2**, 203–218 (1982).

[31] M. Francas & E. C. Larimer, 'Impact of an enhanced electronic messaging system', Paper presented to the *International Communication Association*, May (1984) San Francisco.

[32] F. Korzenny, 'A theory of electronic propinquity: mediated communication in organizations', *Communication Research*, **5**, 3–24 (1978).

[33] L. L. Svenning, 'Individual reponse to an organizationally adopted telecommunications innovation: the difference among attitudes, intentions, and projec-

tions', Paper presented to the *International Communication Association*, May (1985) Honolulu.

[34] E. S. Fredin, 'The dynamics of communication and information in groups: the impact of two-way cable television in organizations', Doctoral dissertation (1980) University of Michigan.

[35] S. R. Hiltz & M. Turoff, 'Structuring computer-mediated communication systems avoid information overload', *Communications of the ACM*, **28**, 680–689 (1985).

[36] J. R. Pierce & E. C. Posner, *Introduction to Communication Science and Systems*, (1980) Plenum, New York.

[37] B. Meek, 'Towards the 21st century', in A. Burns (ed.), *New Information Technology*, pp. 216–243, (1984) Wiley, New York.

[38] F. Williams, *Technology and Communication Behaviour*, (1987) Wadsworth, Belmont.

Index

Index

Mathematics and its Applications

Series Editor: G. M. BELL, Professor of Mathematics, King's College London (KQC), University of London

Statistics and Operational Research

Editor: B. W. CONOLLY, Professor of Operational Research, Queen Mary College, University of London